高等职业院校岗课赛证融通新形态一体化教材

微信小程序项目开发

张严林　王建宣　曾浪勇　主　编
何圣华　黄炳德　赖玉凤　胡迪亮　副主编

電子工業出版社
Publishing House of Electronics Industry
北京·BEIJING

内 容 简 介

本教材内容涵盖了微信小程序项目开发所需要的大部分技术，由 8 个项目 42 个任务组成。除了第一个项目，其余 7 个项目都包含一个完整的微信小程序项目开发案例，这些案例经过了精心制作，贴近企业级应用，部分代码直接来源于编者在企业技术开发中的实际项目。

本教材适合作为高等院校、高等职业院校计算机相关专业的教材，计算机编程爱好者，特别是前端编程爱好者也可以将其作为参考书。本教材是广东机电职业技术学院“嵌入式技术应用”专业省级教学资源库中“跨平台应用开发技术”课程的配套教材。在课程资源库中有各种教学素材，包括教学大纲、PPT、微课视频、程序代码、配套习题、考试题库、教学动画等。读者可扫描书中二维码观看微课视频，也可登录华信教育资源网下载相关资源。

图书在版编目（CIP）数据

微信小程序项目开发 / 张严林，王建宣，曾浪勇主编. -- 北京 : 电子工业出版社，2024. 12. -- ISBN 978-7-121-49436-9

Ⅰ. TN929.53

中国国家版本馆 CIP 数据核字第 2025DA1438 号

责任编辑：王　璐
印　　刷：天津嘉恒印务有限公司
装　　订：天津嘉恒印务有限公司
出版发行：电子工业出版社
　　　　　北京市海淀区万寿路 173 信箱　　邮编：100036
开　　本：787×1092　1/16　印张：15.75　　字数：403 千字
版　　次：2024 年 12 月第 1 版
印　　次：2024 年 12 月第 1 次印刷
定　　价：53.00 元

凡所购买电子工业出版社图书有缺损问题，请向购买书店调换。若书店售缺，请与本社发行部联系，联系及邮购电话：（010）88254888，88258888。

质量投诉请发邮件至 zlts@phei.com.cn，盗版侵权举报请发邮件至 dbqq@phei.com.cn。

本书咨询联系方式：（010）88254580，zuoya@phei.com.cn。

前　言

微信小程序是目前非常热门的轻量级应用，用户只要使用微信“扫一扫”或“搜一搜”即可打开，操作简便，真正实现了各种应用“触手可及”。学习微信小程序开发并不难，微信官方也提供了大量的文档资料，但是初学者只靠自学官方文档是不够的，因为实际的开发需求往往非常复杂，关键在于如何找到正确的思路和解决方案。因此，只有积累大量的实践经验，才能高效地完成开发工作。本教材围绕微信小程序项目开发的学习，将微信小程序项目开发的技术内容融入 7 个实用的企业应用案例中。通过学习实际的企业案例，培养学生的学习兴趣和动手能力。

党的二十大报告指出，加快发展数字经济，促进数字经济和实体经济深度融合，打造具有国际竞争力的数字产业集群。在这一时代浪潮下，本教材应运而生。微信小程序作为连接线上线下、赋能各行各业的创新应用，已成为数字经济发展的重要组成部分。

1．编写理念

本着落实立德树人、强化育人导向的原则，坚持德技并修、三全育人的教育理念，以学生为核心，以能力为本位，突出职业教育的类型特色，为配合广东省省级教学资源库课程“跨平台应用开发技术”，实现线上、线下的混合教学模式，编者编写了这本教材。本教材采用项目任务形式，紧扣以学生为中心的主线，在教材中融入课程思政元素，全面培养学生的爱党爱国情怀、爱岗敬业精神和成为高素质技术技能人才的能力。

2．内容设计

本教材由 8 个项目 42 个任务组成。除了第一个项目，其余 7 个项目都包含一个完整的微信小程序项目开发案例，紧贴企业实际需求，融合几乎所有常用的微信小程序开发技术。每个项目再由若干个任务组成，这些任务有的是重点知识点的介绍，但大部分任务是围绕企业案例展开的。

以项目三“天气预报”为例，该项目通过查询外部网站实现实时天气、24 小时天气预报和 7 天天气预报的显示。类似这种查询外部网站得到信息并显示在微信小程序界面中，正是企业常见需求。项目三由 3 个任务组成，前两个任务介绍主要知识点 Picker 组件和 wx.request 的用法，任务三是制作天气预报小程序。

本教材还在任务中融入课程思政元素，以培养学生的爱国精神和民族自豪感。

3．教材特点

1）精心制作的企业案例贯穿整本教材

编者拥有 10 多年的企业工作经验和 10 多年的高职教学经验，在教学过程中也承担了多项企业技术开发项目，其中不乏微信小程序开发等移动应用开发项目，因此在教学中始终紧贴企业实际需求。本教材不仅仅讲授知识点，而且将知识点融入 7 个贴近企业实际需求的完整案例中。这些案例经过了精心制作，部分代码直接来源于编者在企业技术开发中的实际项目。

2）将思政元素融入工作任务中

结合每个项目的业务特点，有机融入课程思政元素。本教材将“思政育人”的基本精神与项目任务有机结合起来，秉承“能融则融、宜融尽融、以德促学、以学彰德”的基本原则，培养具备爱国情怀与社会责任感的文化传播者。

3）依托省级教学资源库，教学资源丰富

本教材是广东机电职业技术学院“嵌入式技术应用”专业省级教学资源库中“跨平台应用开发技术”课程的配套教材。在课程资源库中有各种教学素材，包括教学大纲、PPT、微课视频、程序代码、配套习题、考试题库、教学动画等。读者可扫描书中二维码观看微课视频，也可登录华信教育资源网下载相关资源。

4．编写团队

编写团队中的多名教师都有 10 多年的企业工作经验，有共同承担企业技术开发项目的经历，此外，还有 4 名企业人员参与编写，因此在编写过程中融合企业实际需求的理念高度一致。

本教材由广东机电职业技术学院张严林教授、王建宣副教授、曾浪勇老师主编。张严林教授是广东省高等学校“千百十工程”省级培养对象、广东省技术能手。张严林负责编写项目二的任务一、项目三的任务三、项目四的任务三到任务六、项目五的任务五到任务十一、项目六的任务五和任务六、项目七、项目八的任务二到任务八，以及习题及所有思政课堂；王建宣负责编写项目一、项目二的任务二和任务三、项目五的任务一到任务四、项目八的任务一；曾浪勇负责项目策划并提供所需的各种素材；何圣华副教授负责编写项目三的任务一和任务二、项目四的任务一和任务二、项目六的任务一到任务四；黄炳德、赖玉凤、胡迪亮协助项目策划并编写了部分项目代码。

5．建议学时

本教材建议教学学时为 60 学时，每个项目的建议学时数分别是 2、6、6、8、16、6、6、10。

因编者水平有限，书中难免会有疏漏之处，敬请各位专家、读者批评指正。

编　者

目　录

项目一　微信小程序与开发环境

2022 年 11 月 16 日，腾讯公司公布的 2022 年第三季度的财报显示，微信月活跃用户达到 13.09 亿，同比增长 3.7%。而且，微信小程序服务了更多商业与民生服务应用场景。微信小程序日活跃用户数突破 6 亿，同比增长超 30%；日均使用次数实现了更快增长，同比增长超 50%。微信加深了微信小程序在食品、服装等主要行业的应用，越来越多的线下商户将其会员与积分系统同微信小程序结合，并通过微信小程序建立多渠道零售体系。此外，微信小程序还积极助力零售企业融合实体业务与数字化能力，实现逆势增长。其中，购物中心、百货行业积极开拓微信小程序收银和数字化会员，发力“线上+线下”全域经营，该行业微信小程序交易额同比增长超 70%。

为什么微信小程序的日活跃用户数能突破 6 亿这个惊人的数量呢？这是由微信小程序的优点决定的。与传统的原生 App 相比，微信小程序有以下优点：

（1）无须安装。微信小程序可以直接在微信中打开，通过搜一搜、扫码、朋友推荐等数十种方式获取，无须安装和卸载，用户体验极佳。

（2）不占手机存储空间。微信小程序基本只是一个用户的操作入口界面，不占手机存储空间，数据基本存储在云端，而实现的功能完全不输原生 App。

（3）开发成本低。微信小程序的开发过程类似于网页开发，如果学过 JavaScript、CSS 3 和 HTML 5，只需再结合微信 API 就可以快速开发出功能完善的微信小程序。微信 API 是微信团队重点打造的开发平台，包含移动开发的方方面面。比如，手机的所有硬件设备都可以通过微信 API 方便地访问。因此，微信小程序可以完全实现原生 App 能实现的所有功能，而且开发过程远比原生 App 简单。

（4）用户无须登录微信小程序。只要用户登录了微信，微信小程序就可以通过授权获取用户微信中的个人信息，无须用户登录各个微信小程序，极大地方便了用户，提高了用户的使用黏性。

（5）可以实现跨平台访问。开发原生 App 必须同时开发 Android 和 iOS 平台，而微信小程序借助微信，无须做任何额外工作就具备了跨平台特性。

微信小程序的优点还有很多，从近年就业市场的招聘情况来看，微信小程序工程师的招聘及薪资待遇呈走高趋势。

下面开始微信小程序学习入门之旅。

任务一　搭建开发环境

1.1.1　注册账号

https://mp.weixin.qq.com/是微信公众平台的网址，在后面的学习中经常要使用到，建议把它加入浏览器的收藏夹。

如图 1.1.1 所示，进入微信公众平台后，先将鼠标指针划过右下角的“小程序”，单击“查看详情”按钮，在弹出的页面中单击“前往注册”按钮，如图 1.1.2 所示，再按照提示一步一步地完成实名注册即可。

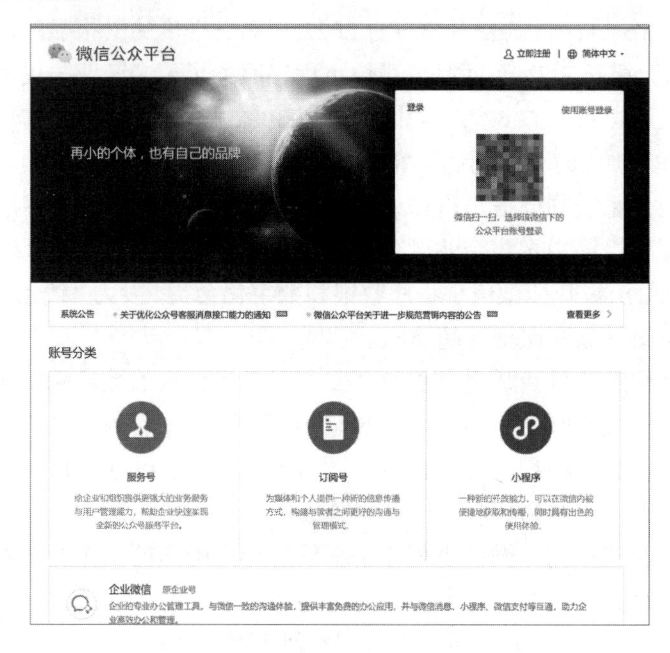

图 1.1.1　微信公众平台

图 1.1.2　注册页面

注册成功后，再次进入微信公众平台，使用微信扫码登录。登录后先在界面左侧选择“开发|开发管理”，再选择“开发设置”，如图 1.1.3 所示，可以看到已经注册的 AppID，在新建小程序项目时会使用到它。

图 1.1.3 获取 AppID

1.1.2 下载并安装微信开发者工具

在图 1.1.2 所示页面中滚动鼠标到页面底部，看到如图 1.1.4 所示页面，单击“开发者工具”，进入如图 1.1.5 所示页面，单击“微信开发者工具”链接，在弹出的页面中下载适合自己电脑的、最新的稳定版本。

图 1.1.4 微信开发者工具

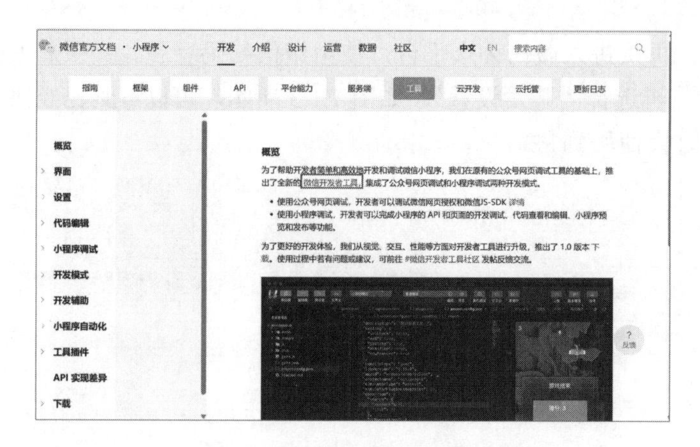

图 1.1.5　下载微信开发者工具

运行程序，按照提示一步步安装即可。

需要说明的是，图 1.1.5 所示页面是微信官方文档，内容详尽。单击“指南”按钮可以进行学习；单击“框架”按钮可以查看微信小程序的框架配置、框架接口、WXML 和 WXS 等参考文档；单击“组件”按钮可以查看微信小程序用到的各种组件；单击“API”按钮可以查看微信平台提供的各种 API 的详细说明，等等。总之，微信官方文档是开发微信小程序的重要助手。

任务二　创建第一个微信小程序

1.2.1　第一个微信小程序

下面以版本 1.06.2306020 win32-x64 为例来演示如何创建微信小程序。

第一次运行微信开发者工具需要用微信扫码登录微信开发平台。登录后新建项目，如图 1.2.1 所示。

图 1.2.1　创建小程序

可以先选择目录，则项目名称自动以目录名称命名，也可以修改项目名称。AppID 可以使用测试号，也可以使用已注册的 AppID。本教材不学习云服务，新建的所有项目都选择“不使用云服务”。选择“不使用模板”，单击“确定”按钮完成新建项目。微信开发者工具主界面如图 1.2.2 所示。

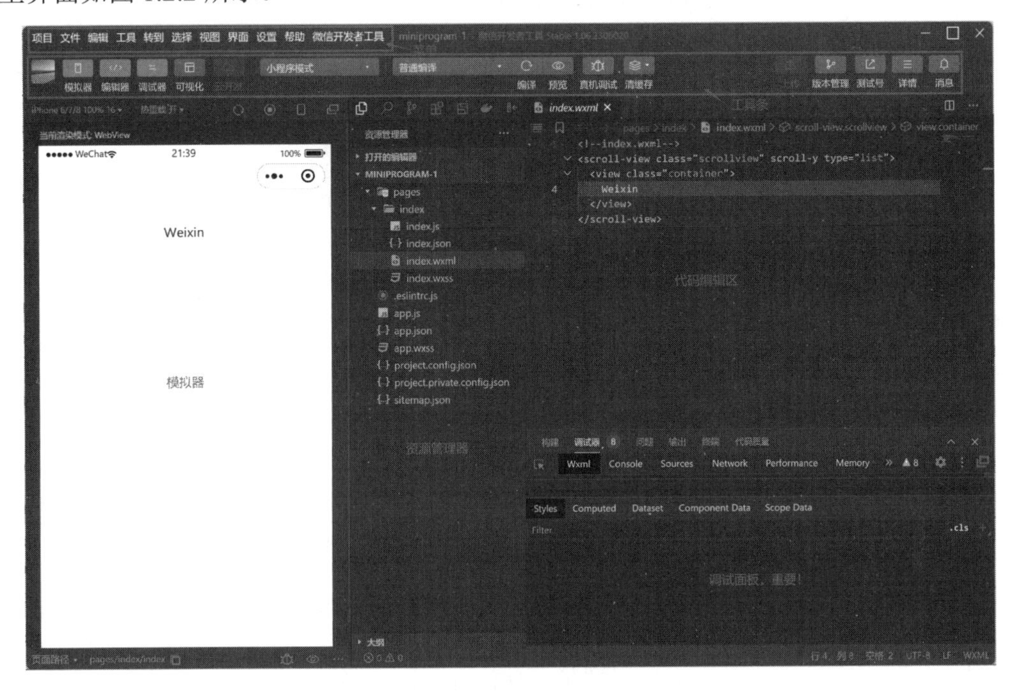

图 1.2.2　微信开发者工具主界面

微信开发者工具与常见的编程工具类似，显著的不同就是其左侧的模拟器，中间是资源管理器，右侧是代码编辑区，顶部是常用菜单和工具条，右下角是调试面板。

在资源管理器中展开“pages/index”，单击 index.wxml 文件，在代码编辑区将“Weixin”修改为“Hello World！”，单击工具条上的“编译”按钮，模拟器将显示“Hello World！”。单击工具条上的“预览”按钮，使用微信扫码，则实现了在手机上运行这个小程序。第一个微信小程序就完成了。

1.2.2　调试面板

调试面板在调试代码过程中经常需要使用到，下面对其做简要说明，如表 1.2.1 所示。

表 1.2.1　调试面板说明

名称	功能	说明
Wxml	页面结构	显示渲染后的页面布局及样式，在调试页面结构及样式时经常使用。当发现页面没有按照设计正常显示时，通常使用此面板进行调试。可以根据页面显示的内容快速对应到 wxml 代码，也可以通过实时修改样式来查看效果

续表

名称	功能	说明
Console	控制台	显示调试信息和各种错误信息。在编写代码过程中，控制台会实时显示各种错误信息，帮助用户及时发现错误并改正；也可以编写调试代码调试程序。可以在此面板执行指令和代码，实时查看结果
Sources	源代码	查看编译处理后的脚本文件
Network	网络	调试网络请求（wx.request）时使用，可以看到调用的服务器的各种资源，以及服务器返回的各种信息
Performance	性能	查看逻辑层的执行情况
Memory	内存	分析内存使用情况，检查有没有出现内存泄露
AppData	应用数据	查看运行中的数据，并且可以修改、查看实时效果
Storage	存储	查看调用 wx.setStorage 或 wx.setStorageSync 后的数据存储情况，还可以进行删除、修改、新增操作
Security	安全	调试当前程序的安全和认证等问题，并且确保已经在程序上正确使用 HTTPS
Sensor	传感器	目前仅支持地理位置和加速度传感器，可以模拟这两种传感器的数据，用于调试程序
Mock	模拟	可以模拟部分 API 的调用结果
Audits	审计	在运行过程中进行实时检查，定位出可能导致体验不好的地方，并给出优化建议
Vulnerability	漏洞	发现并修复小程序内的接口安全漏洞，提升安全性

下面以常用的 Console 面板为例进行用法说明。打开 index.wxml 文件，先故意造成一个错误，删除第三行最后的“>”字符，然后单击工具条上的“编译”按钮，选择调试面板中的 Console 面板，可以看到显示编译错误，如图 1.2.3 所示。

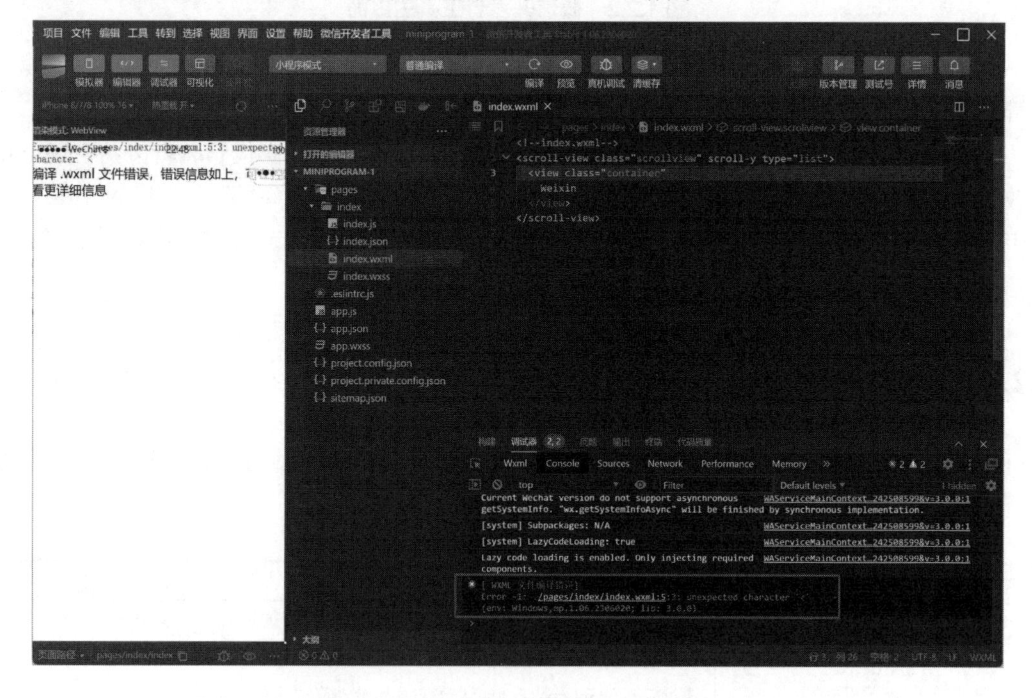

图 1.2.3　Console 面板显示错误信息

再补回“>”字符，打开 index.js 文件，将光标定位在第二行的{}中，按回车键，输入以下代码：

```
1      onLoad: function () {
2        console.log('Hello World!')
3      }
```

console.log 方法用于在控制台打印调试信息。单击工具条上的“编译”按钮，Console 面板会显示调试信息，如图 1.2.4 所示。调试信息的右边显示的是打印该信息的 js 文件的位置。

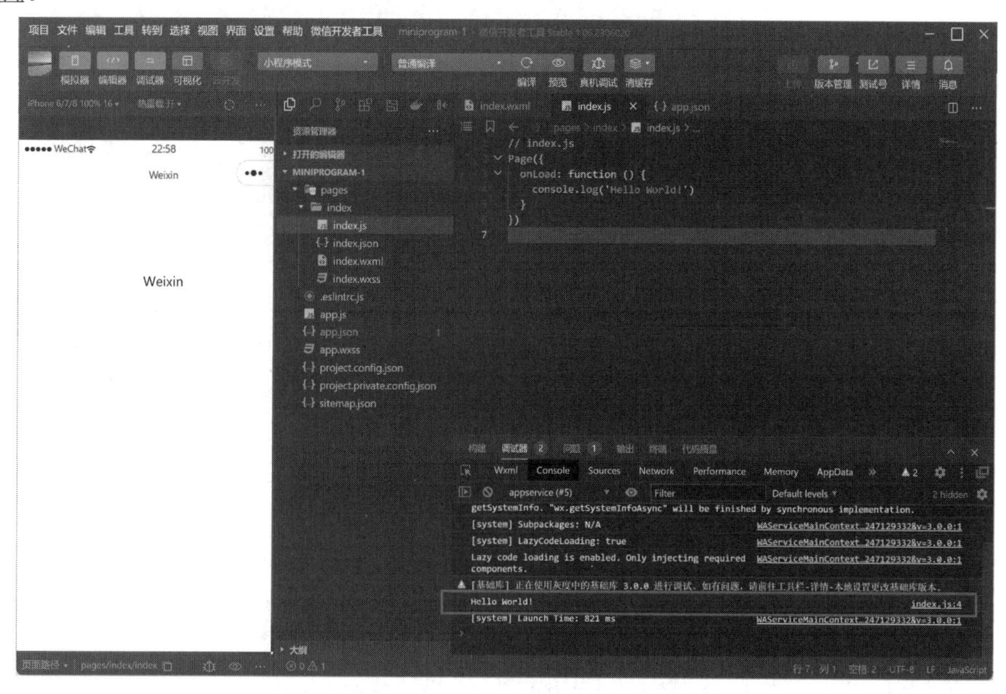

图 1.2.4 Console 面板显示调试信息

1.2.3 格式化代码

创建微信小程序需要使用 js 代码、类似 HTML 的 wxml 代码及与 CSS3 几乎完全一样的 wxss 代码。初学者编写代码时经常不缩进或缩进不规范，造成层次关系不清晰，调试困难，因此代码排列整齐对调试非常重要。微信开发者工具提供了优秀的格式化代码的工具。

建议代码统一采用两个空格代替 Tab 缩进的排版方式，这样代码更紧凑，在各种文本编辑器中查看的代码也基本一致。

打开项目根文件夹下的 project.config.json 文件，修改 editorSetting 的设置，如图 1.2.5 所示，将代码改成：

```
1      "editorSetting": {
2        "tabIndent": "insertSpaces",
```

```
3           "tabSize": 2
4         }
```

这样代码就会使用两个空格代替 Tab 缩进。

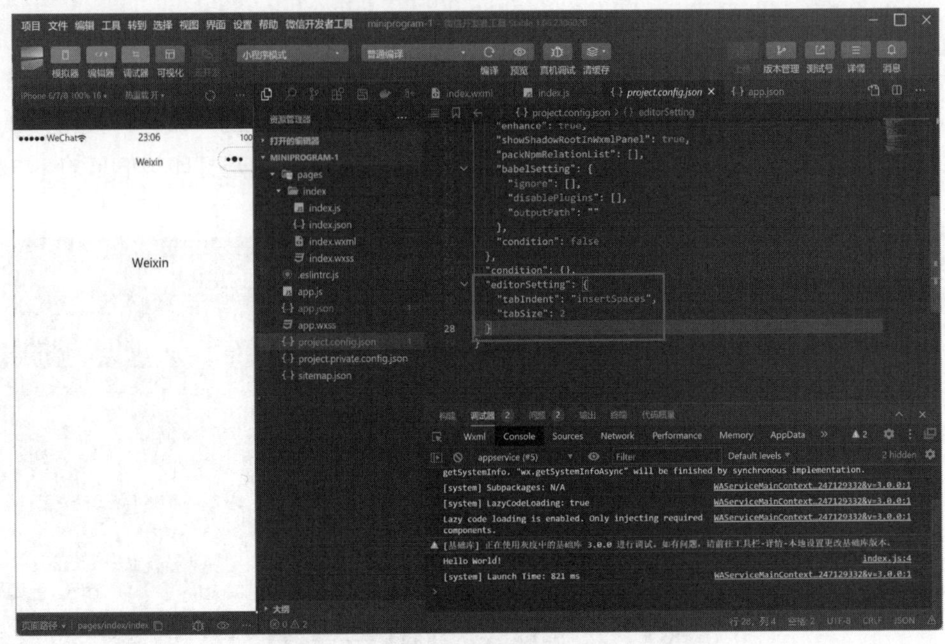

图 1.2.5　设置使用两个空格代替 Tab 缩进

在编写代码过程中，可以随时右击，在弹出的快捷菜单中选择“格式化文档”或“格式化文档，方法是使用...”命令，如图 1.2.6 所示。

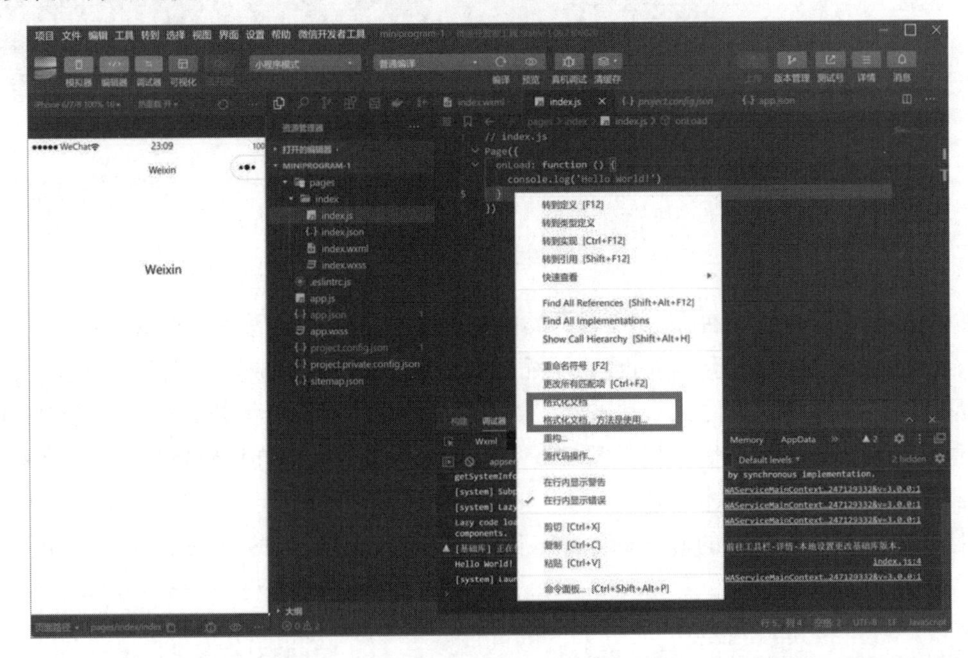

图 1.2.6　格式化文档

第一次选择“格式化文档”命令时会让用户选择格式化方法，建议选择默认值，如图 1.2.7 所示，则以后选择“格式化文档”命令时都会使用该默认值。如果需要修改格式化方法，就选择“格式化文档，方法是使用...”命令。

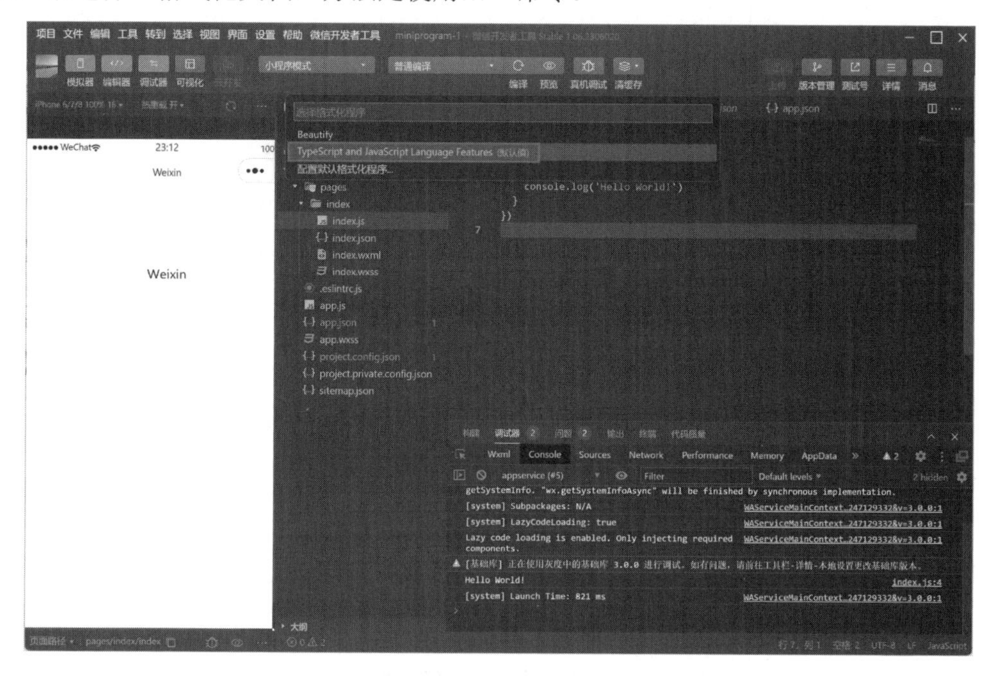

图 1.2.7　格式化的方法

对微信小程序中使用的所有文件都可以按照此方法进行格式化，这样代码更整齐、美观，而且调试方便。

任务三　了解微信小程序代码组成

1.3.1　项目文件

每个微信小程序项目都有固定的项目文件，如表 1.3.1 所示。

表 1.3.1　项目文件用途说明

项目文件	用途说明
.eslintrc.js	配置 Eslint 文件，需要先安装 Eslint。Eslint 用来规范代码书写，使不同的开发者编写的代码风格统一
app.js	微信小程序的入口文件，掌控整个小程序的生命周期，一些全局的属性、变量也存放在这个文件中
app.json	微信小程序的全局配置，包括所有页面路径、界面表现、网络超时时间、底部 tab 等
app.wxss	全局样式，所有页面都可以使用

续表

项目文件	用途
project.config.json	在使用一个工具时，通常用户会根据喜好进行个性化配置，如界面颜色、编译配置等。 考虑到这一点，微信开发者工具在每个项目的根目录都会生成一个 project.config.json 文件，用户在工具上做的任何配置都会被写入这个文件。当用户重新安装工具或更换电脑工作时，只要载入同一个项目的代码包，微信开发者工具就会自动恢复到开发项目时的个性化配置，包括编辑器的颜色、代码上传时自动压缩等一系列选项
project.private.config.json	项目私有配置文件。此文件中的内容将覆盖 project.config.json 文件中的相同字段，项目的改动优先同步到此文件中
sitemap.json	微信现已开放小程序内搜索，开发者可以通过 sitemap.json 配置，或者管理后台页面收录开关来配置其小程序页面是否允许微信索引。当开发者允许微信索引时，微信会通过爬虫的形式为小程序的页面内容建立索引。当用户的搜索词条触发该索引时，小程序的页面将可能展示在搜索结果中

1.3.2 页面文件

微信小程序有一个或多个页面，页面文件全部放在/pages/文件夹下。通常每个页面有一个单独的文件夹，每个页面至少包含四类文件（可以根据需要添加），如表 1.3.2 所示。

表 1.3.2 微信小程序页面包含的四类文件

文件	用途
WXML 文件	页面布局文件
WXSS 文件	页面样式文件
JS 文件	页面脚本逻辑文件，可以调用小程序提供的丰富的 API，利用 API 可以很方便地调用微信提供的功能，例如，获取用户信息、本地存储、微信支付等
JSON 文件	页面配置文件

1.3.3 四类文件简介

1. WXML 文件

WXML 由标签、属性等构成，和 HTML 类似，但是也有不同，下面一一阐述。

1）常用标签名称不同

编写 HTML 时经常使用的标签有 div、p 和 span 等，开发者在编写一个页面时，可以根据这些基础标签组合出不一样的组件，如日历、弹窗等。既然大家都需要这些组件，为什么不把它们包装起来重复使用，以便提高开发效率。

小程序的 WXML 常用的标签有 view、button、text 等，它们是小程序为开发者包装好的基本能力，小程序还提供了地图、视频、音频等组件能力。

2）增加 wx:if 等属性及{{ }}等表达式

在网页的一般开发流程中，通常通过 JS 操控 DOM（对应 HTML 的描述产生的树），以引起界面的一些变化，响应用户的行为。例如，当用户单击某个按钮时，JS 会记录一些状态到 JS 变量中，同时通过 DOM API 操控 DOM 的属性或行为，进而引起界面的某些变化。当项目越来越大时，代码会充斥非常多的界面交互逻辑和程序的各种状态变量，显然这不是一种很好的开发模式。因此就有了 MVVM 开发模式（如 React 和 Vue），它提倡把渲染和逻辑分离。简单来说，就是不让 JS 直接操控 DOM，它只需要管理状态，并通过模板语法来描述状态和界面结构的关系即可。

小程序的框架也使用了这种思路，例如，需要把“Hello World”字符串显示在界面上，WXML 是这样写的：

```
<text>{{msg}}</text>
```

而 JS 只需要管理状态即可：

```
this.setData({ msg: "Hello World" })
```

通过{{ }}把一个变量绑定到界面上，我们称为数据绑定。仅仅通过数据绑定还不能够完整地描述状态和界面结构的关系，还需要 if/else、for 等控制能力。在小程序中，这些控制能力都使用以 wx:开头的属性来表达。

2. WXSS 文件

WXSS 具有 CSS 大部分的特性，小程序在 WXSS 上也做了扩充和修改。

1）新增尺寸单位

在编写 CSS 样式时，开发者需要考虑手机设备的屏幕有不同的宽度和设备像素比，采用一些技巧来换算尺寸单位。而 WXSS 在底层支持新的尺寸单位 rpx，开发者可以免去单位换算的烦恼，只要交给小程序底层即可。因为单位换算采用浮点数运算，所以运算结果和预期结果会有一点点偏差。

2）提供全局样式和局部样式

开发者可以编写一个 app.wxss 作为全局样式，其作用于当前小程序的所有页面，局部页面样式 page.wxss 仅对当前页面生效。

此外，WXSS 仅支持部分 CSS 选择器。

3. JS 文件

一个服务只有界面展示是不够的，还需要与用户交互：响应用户的点击、获取用户的位置等。在小程序中，开发者通过编写 JS 脚本文件来处理用户的操作：

```
<view>{{ msg }}</view>
<button bindtap="clickMe">点击我</button>
```

点击 button 按钮时，如果希望界面上的“msg”显示成“Hello World”，就在 button 按钮上声明一个属性 bindtap，在 JS 文件中声明 clickMe 方法来响应这次点击操作：

```
Page({
  clickMe: function() {
    this.setData({ msg: "Hello World" })
  }
})
```

响应用户的操作就是这么简单。

4．JSON 文件

JSON 文件被包裹在一个大括号中{}，通过 key-value 的方式表示数据，而且 key 值必须包裹在一对双引号或单引号中。在编写 JSON 文件时，忘记为 key 值加双引号或单引号是常见错误。

JSON 的值只能是以下几种数据格式，其他任何格式都会触发报错，如 JavaScript 中的 undefined。

- 数字，包含浮点数和整数。
- 字符串，包裹在双引号中。
- Bool 值，true 或 false。
- 数组，包裹在中括号中。
- 对象，包裹在大括号中。
- Null。

在 JSON 文件中无法使用注释，试图添加注释将触发报错。

1.3.4 微信小程序协同工作与发布

开发一个大型的小程序需要团队协作，一般开发团队由项目管理人员管理，包括产品组、设计组、开发组和测试组。

在小程序的后台可以进行项目开发组的成员管理。进入微信公众平台，使用微信扫码登录。登录后在界面左侧选择“管理|成员管理”，如图 1.3.1 所示，在此可以编辑或新增成员，还可以为成员赋予不同的角色。

小程序开发分为需求分析、设计、开发、体验、测试和发布，与常规软件开发相比，其有一个独特的发布步骤。一个小程序从开发完成到上线一般要经过预览→上传代码→提交审核→发布等步骤，相应地，小程序的版本也分为开发版本、体验版本、审核中版本、线上版本。

小程序开发完成后，先单击工具条上的“上传”按钮上传代码，然后进入微信公众平台。登录后在界面左侧选择“管理|版本管理”，如图 1.3.2 所示。如果有开发版本，会出现“提交审核”按钮，单击此按钮，开发版本就被提交到微信后台审核。如果审核通过，开发版本就成为线上版本。线上版本有一个圆形的小程序码，任何人都可以扫此码访问小程序。

如果审核不通过，就会收到邮件通知，详细说明不通过的原因，修改后可以再次上传、提交审核。

图 1.3.1　成员管理

图 1.3.2　版本管理

项目小结

本项目首先介绍了微信小程序的特点，搭建了微信小程序的开发环境，并创建了第一个微信小程序，最后简要介绍了微信小程序代码的组成、微信小程序协同工作与发布。

习题

一、判断题

1．微信小程序的全局配置文件是 project.config.json。（　　）

2．微信小程序的页面布局文件是 WXML 文件。（　　）

3．微信小程序发布成功后才能成为线上版本。（　　）

二、选择题

1．（　　）不是微信小程序的特点。

A．无须安装　　B．开发成本低

C．可以替代 App　　D．可以实现跨平台访问

2．显示调试信息、各种错误信息的面板是（　　）。

A．Wxml　　B．Console

C．Sources　　D．Audits

3．微信小程序的入口文件是（　　）。

A．app.js　　B．app.json

C．app.wxss　　D．app.wxml

三、填空题

1．开发一个大型的小程序需要团队协作，一般开发团队由项目管理人员管理，包括产品组、设计组、开发组和___________。

2．小程序开发分为需求分析、设计、开发、体验、测试、发布等流程，与常规软件开发相比，其有一个独特的____________步骤。

3．微信小程序页面样式文件是_________文件。

项目二　答题小程序

任务一　制作答题小程序

本项目创建了一个简单、完整的答题小程序，重要知识点在任务二和任务三讲解。

2.1.1　项目介绍

答题小程序是一个简单的判断题答题并计算得分的小程序，包括六道判断题，可以选择“正确”或“错误”选项进行答题；点击“下一题”按钮跳转到下一题；点击“上一题”按钮跳转到上一题；点击“交卷”按钮完成答题，并显示得分；点击“重做”按钮重新答题，如图 2.1.1 所示。

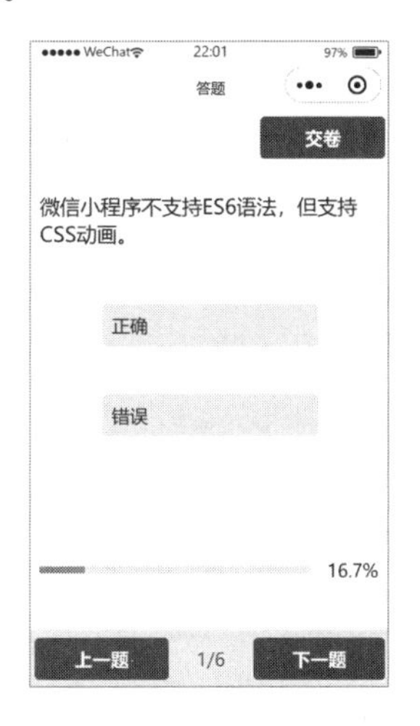

图 2.1.1　答题小程序运行结果

扫一扫

微课：答题项目操作

2.1.2　创建项目

打开微信开发者工具，新建项目，名称为 quiz，使用测试账号，不使用模板。

打开 app.json，修改“navigationBarTitleText”为“答题”。

复制资源文件夹中 quiz 下面的 images 文件夹到新建的 quiz 微信小程序文件夹，使该文件夹与 pages 文件夹并列。images 文件夹中有“正确.png”和“错误.png”两个图片文件。这两张图片用于显示答题结果。

2.1.3 设计答题界面

打开 index.wxml 文件，删除原来的代码，输入以下代码：

```
<view class="finish" bindtap="finish">交卷</view>
<view class="question">{{title}}</view>
<view class="result" bindtap="yes">正确</view>
<view class="result" bindtap="no">错误</view>
<view class="toolbar">
  <view bindtap="prev">上一题</view>
  <text class="progress_text">{{progress_text}}</text>
  <view bindtap="next">下一题</view>
</view>
<progr ess percent="{{percent}}" show-info="true"></progress>
```

打开 index.wxss 文件，输入以下代码：

```
.finish {
  text-align: center;
  margin: 10rpx 10rpx 10rpx 480rpx;
  padding: 20rpx;
  color: #fff;
  font-weight: bold;
  background: darkgreen;
  border-radius: 10rpx;
}

.toolbar {
  display: flex;
  align-items: center;
  position: fixed;
  bottom: 0rpx;
  width: 100vw;
  height: 120rpx;
  background: #eee;
}

.toolbar>view {
  flex: 1;
  text-align: center;
```

```
24      margin: 10rpx;
25      padding: 20rpx;
26      color: #fff;
27      font-weight: bold;
28      background: darkgreen;
29      border-radius: 10rpx;
30    }
31
32    .progress_text {
33      color: darkgreen;
34      background: #eee;
35      margin: 0 50rpx;
36    }
37
38    .question {
39      margin: 50rpx auto;
40      padding: 20rpx;
41      font-size: large;
42    }
43
44    .result {
45      margin: 100rpx 150rpx;
46      padding: 20rpx;
47      background: #eee;
48      border-radius: 10rpx;
49    }
50
51    .result.active {
52      background: lightgreen;
53    }
54
55    progress {
56      position: fixed;
57      width: 710rpx;
58      bottom: 200rpx;
59      margin: 20rpx;
60    }
61
62    button {
63      background-color: darkgreen;
64      color: #fff;
65    }
```

在本段代码中，第 12 行代表 toolbar 采用 Flex 布局，默认方向是横向；第 13 行代表居中对齐；第 14 行代表固定定位；第 15 行代表对齐底部。也就是说，toolbar 固定在底部，采用 Flex 布局。Flex 布局和尺寸单位 rpx 将在本项目任务二中介绍。

2.1.4 设计逻辑代码

微课：答题项目布局代码

1）设计数据

打开 index.js 文件，将光标定位于 Page({})的大括号中间，按回车键，输入以下代码：

```
1      data: {
2        title: '',
3        finish: false,
4        questions: [{
5          title: '微信小程序不支持 ES6 语法，但支持 CSS 动画。',
6          answer: false
7        }, {
8          title: '在微信小程序中，AppID 又称为小程序 ID，是每个小程序的唯一标识。',
9          answer: true
10       }, {
11         title: '微信小程序能够实现复杂的应用，将取代原生 App。',
12         answer: false
13       }, {
14         title: '微信小程序运行环境是微信客户端，可以实现跨平台。',
15         answer: true
16       }, {
17         title: '微信小程序开发类似于传统的网页开发，微信内部对语言进行了定制。',
18         answer: true
19       }, {
20         title: '微信公众号就是微信小程序账号，只有通过注册才可进行微信小程序的开发。',
21         answer: false
22       }
23       ],
24       progress_text: '',
25       result: null,
26       score: 0
27     },
```

其中，答题放在第 4 行的数组 questions 中，该数组目前有 6 个元素，代表 6 道判断题。title 代表题目、answer 代表答案，后面还会增加 submit 和 result 两个属性，分别代表是否提交了答案，以及提交的答案。

代码中的其他属性都代表在界面显示的相关信息，如 finish 代表是否交卷、score 代表得分，等等。

在此需要特别注意的是，只有放在 data{}中的数据，在用户界面（wxml 代码）才能被看到，后面会重点讲解。

微课：答题项目 js 代码 1

2）设计关键方法 setIndex

为了能够前后移动题目，也为了能够刷新程序的状态，设计了一个关键方法 setIndex。在上面 data{}代码的后面插入以下代码：

```
1        index: 0, //questions 的下标
2        setIndex: function () {
3          //this.setData 更新数据并刷新页面显示
4          this.setData({
5            title: this.data.questions[this.index].title,
6            //toFixed() 将数字转换为字符串，四舍五入到指定的小数位数
7            percent: ((this.index + 1) * 100.0 / this.data.questions.
   length).toFixed(1),
8            progress_text: (this.index + 1) + '/' + this.data.questions.
   length,
9            result: this.data.questions[this.index].submit ?
10             this.data.questions[this.index].result : null
11         })
12       },
```

因页面宽度不够造成的换行，如第 7 行、第 8 行，在输入时不需要换行。

第 4 行中的 setData 方法是微信小程序开发中一个非常重要的方法，用于将数据从逻辑层发送到视图层。通俗来说，就是把数据从逻辑代码（js 代码）发送到界面代码（wxml 代码）。setData 同时做了更新数据和刷新页面显示两件事，更新数据即更新了 this.data 中的数据，也就是上面在 data{}中输入的数据。

本段代码更新了题目、百分比、进度条文字和答题结果等数据，并刷新了页面显示。

扫一扫

微课：答题项目 js 代码 2

3）完成答题代码

在 setIndex 方法后面插入以下代码：

```
1        onLoad: function (options) {
2          this.setIndex()
3        },
4        prev: function (e) {
5          if (this.index > 0) this.index--
6          this.setIndex()
7        },
8        next: function (e) {
9          if (this.index < this.data.questions.length - 1) this.index++
10         this.setIndex()
11       },
12       setResult: function (result) {
13         this.data.questions[this.index].submit = true //submit 代表已提交答案
14         this.data.questions[this.index].result = result //result 代表提交
   的答案
15         this.setData({ result: result })
```

```
16        },
17        yes: function (e) {
18          this.setResult(true)
19        },
20        no: function (e) {
21          this.setResult(false)
22        },
```

onLoad 方法属于页面的生命周期回调，在页面加载时触发。

答题不仅需要处理“上一题”和“下一题”按钮，还需要处理用户的选择——“正确”或“错误”，prev、next、yes、no 分别完成这四项功能。

为了高亮显示用户的选择，在 index.wxss 文件中已经输入了.result.active 的样式——亮绿色背景，现在需要在 WXML 文件中根据用户的选择来调用。

打开 index.wxml 文件，找到以下代码：

```
<view class="result" bindtap="yes">正确</view>
<view class="result" bindtap="no">错误</view>
```

添加粗斜体部分：

```
<view class="result {{result!=null && result? 'active':''}}" bindtap=
"yes">正确</view>
<view class="result {{result!=null && !result? 'active':''}}" bindtap=
"no">错误</view>
```

至此，完成了答题代码，单击微信开发者工具上的“编译”按钮，会自动保存代码并在模拟器中运行，现在可以切换题目并答题了。

4）显示答题结果

打开 index.wxml 文件，先将原有代码用以下标记包围起来：

```
<block wx:if="{{!finish}}">
......
</block>
```

再补充代码，最终代码如下：

```
1      <!-- 条件渲染 -->
2      <block wx:if="{{!finish}}">
3        <view class="finish" bindtap="finish">交卷</view>
4        <view class="question">{{title}}</view>
5        <view class="result {{result!=null && result? 'active':''}}"
       bindtap="yes">正确</view>
6        <view class="result {{result!=null && !result? 'active':''}}"
       bindtap="no">错误</view>
7        <view class="toolbar">
8          <view bindtap="prev">上一题</view>
9          <text class="progress_text">{{progress_text}}</text>
```

```
10        <view bindtap="next">下一题</view>
11      </view>
12      <progress percent="{{percent}}" show-info="true"></progress>
13    </block>
14    <block wx:else>
15      <view class="score">您的得分: {{score}}</view>
16      <button bindtap="reset">重做</button>
17      <!-- 列表渲染 -->
18      <view class="list" wx:for="{{questions}}" wx:key="title">
19        <view class="question" style="margin: 10rpx auto;">
20          {{item.title}}
21        </view>
22        <image src="/images/{{item.submit && item.answer == item.
result ? '正确': '错误'}}.png" />
23      </view>
24    </block>
```

第 2 行采用条件渲染，当 finish 为 false，即没有完成答题时，显示 block 中的内容。第 14 行代表当 finish 为 true 时，即答题完成后，显示该 block 中的内容。

第 18 行采用列表渲染，对数组 questions 进行列表（循环）渲染，即重复显示<view class="list"> …</view>的内容 *N* 次，*N* 是 questions 数组中元素的个数。

打开 index.wxss 文件，在后面补充以下代码：

```
1     .list {
2       display: flex;
3     }
4
5     .list>view {
6       flex: 1
7     }
8
9     .list>image {
10      width: 60rpx;
11      height: 60rpx;
12      margin: auto 20rpx;
13    }
14
15    .score {
16      margin: 30rpx;
17      padding: 20rpx;
18      text-align: center;
19      font-size: x-large;
20      background-color: lightgreen;
21      color: #000;
22    }
```

打开 index.js 文件，在 Page({})内部补充两个方法，代码如下：

```
1        finish: function (e) {
2          let score = 0
3          //对数组 this.data.questions 进行循环
4          this.data.questions.forEach(element => {
5            //如果已提交结果，且结果正确
6            if (element.submit && element.result == element.answer)
7              score++
8          });
9          //则计算得分（百分制）
10         score = 100.0 * score / this.data.questions.length
11         this.setData({
12           finish: true,
13           questions: this.data.questions,
14           score: score.toFixed(1)
15         })
16       },
17       reset: function (e) {
18         //重置所有数据
19         this.data.questions.forEach(element => {
20           element.submit = false
21           element.result = undefined
22         });
23         this.index = 0
24         this.setIndex()
25         this.setData({
26           finish: false,
27           questions: this.data.questions,
28           result: null,
29           score: 0
30         })
31       }
```

至此，代码编写完成，部分代码的解释见注释。编译后，点击“交卷”按钮可以看到答题结果，点击“重做”按钮可以重新答题。

2.1.5 微信小程序生命周期

微信小程序生命周期一般包括创建、装载、显示、隐藏、唤醒、停止、卸载、销毁等。微信小程序应用和每个页面都有生命周期。

微信小程序应用的生命周期如图 2.1.2 所示。

图 2.1.2 微信小程序应用的生命周期

- 第一次打开小程序时，触发 app.js 文件中的 onLaunch 方法（只触发一次）。
- 初始化完成后，触发 app.js 文件中的 onShow 方法，监听小程序显示。
- 小程序从前台显示进入后台运行，触发 app.js 文件中的 onHide 方法。
- 小程序从后台运行进入前台显示，触发 app.js 文件中的 onShow 方法。
- 小程序在后台运行一定时间（目前是 5 分钟）后，会被微信主动销毁。
- 系统资源占用过多，可能会被系统销毁，或者被微信主动回收。

页面生命周期如图 2.1.3 所示。

- 加载页面时，触发页面的 onLoad 方法，一个页面只调用一次。
- 页面载入后，触发页面的 onShow 方法显示页面，每次打开页面都调用一次。
- 初次渲染页面时，触发页面的 onReady 方法渲染页面元素和样式，一个页面只调用一次。
- 当小程序后台运行或跳转到其他页面时，触发页面的 onHide 方法。
- 当小程序由后台运行进入前台显示或重新进入页面时，触发页面的 onShow 方法。

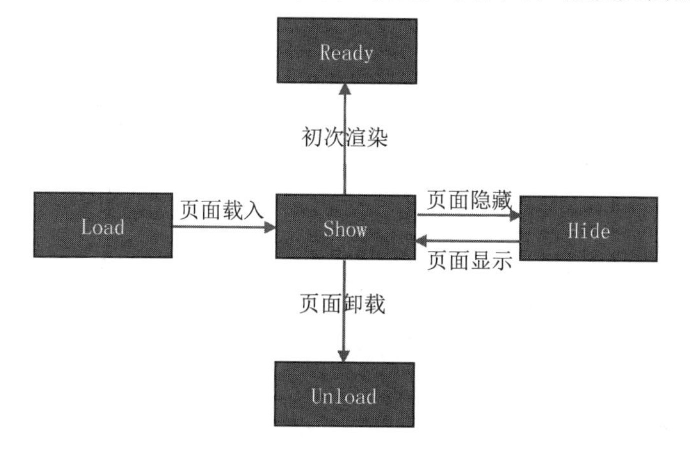

图 2.1.3　页面的生命周期

需要注意的是，页面初次加载运行时，执行的顺序是 onLoad→onShow→onReady，onLoad 和 onReady 在页面的生命周期中只执行一次。

任务二　使用 Flex 布局

Flex 是 Flexible Box 的缩写，Flex 布局又称弹性布局。Flex 布局的代码简单，但功能强大。使用 Flex 布局，可以很容易地制作排列整齐、自动适应屏幕大小的页面，因此其在微信小程序开发中得到了广泛应用。本教材后续的大部分项目都使用了 Flex 布局。

Flex 布局使用简单，只需为某容器组件（通常是 View）添加“display: flex;”样式，再配合少量 Flex 样式即可。Flex 布局常见属性如表 2.2.1 所示。

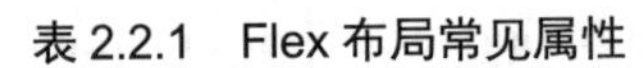

表 2.2.1　Flex 布局常见属性

属性	属性值及含义
flex-direction	• row：默认值，如果不指定 flex-direction，则默认为 row，横向，从左到右排列。 • row-reverse：横向，从右到左排列。 • column：纵向，从上到下排列。 • column-reverse：纵向，从下向上排列。 Flex 布局有“主轴”和“交叉轴”的概念。如果横向排列，则主轴为横向，交叉轴为纵向；否则主轴为纵向，交叉轴为横向
flex-wrap	• nowrap：默认值，不换行，如果当前行或列显示不下就看不到。 • wrap：换行，换行/列显示。 • wrap-reverse：换行，换的行在上面或左面
flex-flow	flex-direction 和 flex-wrap 组合后的简写形式
justify-content	Flex 项目在主轴上的对齐方式。 • flex-start：默认值，左/上对齐。 • flex-end：右/下对齐。 • center：居中对齐。 • space-between：两端对齐，项目之间的间隔相同。 • space-around：每个项目两侧的间隔相等，因此，项目之间的间隔比项目与边框的间隔大一倍
align-items	Flex 项目在交叉轴上的对齐方式。 • flex-start：上/左对齐。 • flex-end：下/右对齐。 • center：居中对齐。 • baseline：第一行文字的基线对齐。 • stretch：默认值，如果项目未设置高度或设置为 auto，就占满整个容器的高度（横向排列）或宽度（纵向排列）
以上属性是作用在父元素（使用 Flex 布局的容器）上的，下面的属性是作用在 Flex 项目上的	
order	项目的排列顺序。数值越小，排列越靠前，默认为 0
flex-grow	定义 Flex 项目的放大比例，默认为 0，即使存在剩余空间，也不放大。 如果所有项目的 flex-grow 属性都为 1，则它们将等分剩余空间（如果有）；如果一个项目的 flex-grow 属性为 2，其他项目的都为 1，则前者占据的剩余空间比其他项目多一倍。 这个属性非常重要，经常应用于让 Flex 项目平均分配或按比例分配剩余空间的情况，适应屏幕大小的同时排列整齐
flex-shrink	定义 Flex 项目的缩小比例，默认为 1，即如果空间不足，则该项目将被缩小。如果所有项目的 flex-shrink 属性都为 1，则当空间不足时，等比例缩小；如果一个项目的 flex-shrink 属性为 0，其他项目的都为 1，则当空间不足时，前者不缩小
flex-basis	flex-basis 属性定义了在分配多余空间之前，项目占据的主轴空间。它的默认值为 auto，即项目的本来大小
flex	flex 属性是 flex-grow、flex-shrink 和 flex-basis 的简写形式，默认值为 0、1 和 auto，后两个属性可选。 通常使用这个属性作为 flex-grow 属性的简写形式，实现 Flex 项目平均分配或按比例分配剩余空间。 推荐使用这种简写形式，本教材后续项目全部采用这种方式

续表

属性	属性值及含义
align-self	align-self 属性允许单个项目有与其他项目不一样的对齐方式，可覆盖 align-items 属性。默认值为 auto，表示继承父元素的 align-items 属性，如果没有父元素，则等同于 stretch。 该属性可能取 6 个值，除了 auto，其他值与 align-items 属性完全一致

2.2.1　测试容器属性

接下来编写代码测试 Flex 布局，可以使用在项目一中创建的第一个小程序，也可以创建一个新的小程序。

1．测试 flex-direction 属性

打开 index.wxml 文件，删除原来的代码，输入以下代码：

```
1    <view class="flex-container">
2      <view>1</view>
3      <view>22</view>
4      <view>333</view>
5      <view>4444</view>
6      <view>55555</view>
7      <view>666666</view>
8    </view>
```

编译后，会看到第 2～7 行的项目上下顺序排列。之所以在 view 中使用不同长度的字符串，是为了后面测试这些 Flex 项目的宽度。

打开 index.wxss 文件，输入以下代码：

```
1    .flex-container {
2      display: flex;
3      background: darkblue;
4      height: 100rpx;
5     }
6
7    .flex-container>view {
8      background: lightgray;
9      margin: 30rpx;
10   }
```

网页设计中的长度单位通常是像素（px），而在微信小程序中经常使用的单位是 rpx，这是因为手机的屏幕大小不同，需要在不同屏幕尺寸下让显示效果类似。

rpx（responsive pixel，响应式像素，或称为逻辑像素）：可以根据屏幕宽度进行自适应。规定屏幕宽度为 750rpx。如在 iPhone6 上，屏幕宽度为 375px，共有 750 个物理像素，则 750rpx = 375px = 750 物理像素，1rpx = 0.5px = 1 物理像素。

第 2 行设置了 Flex 布局，为了更好地查看 Flex 项目的宽度，第 3 行和第 8 行设置了背景色，第 4 行设置了高度，第 9 行设置了外边距。编译后，显示的页面如图 2.2.1 所示。

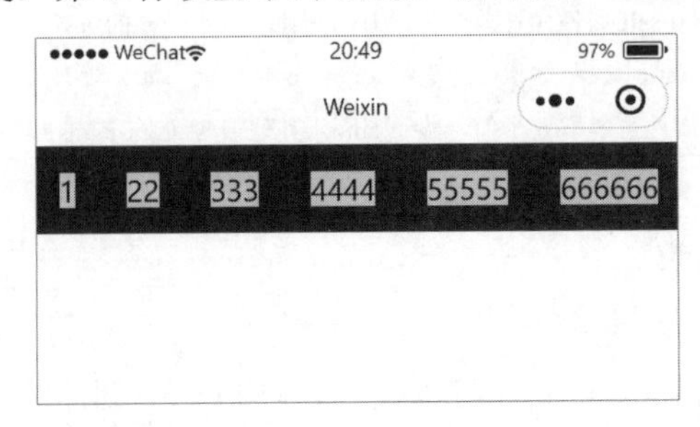

图 2.2.1　Flex 横向布局

可以看到 Flex 项目呈水平横向分布，且宽度是各自的实际宽度。

如果要使 Flex 项目纵向显示，则将.flex-container 的样式改成：

```
1  .flex-container {
2    display: flex;
3    flex-direction: column;
4    background: darkblue;
5    height: 600rpx;
6    width: 200rpx;
7  }
```

第 3 行设置了 Flex 项目纵向排列，而在前面的代码中没有设置这个属性，因此图 2.2.1 是横向排列，因为 flex-direction 属性的默认值是横向排列。此时编译后，显示的页面如图 2.2.2 所示。

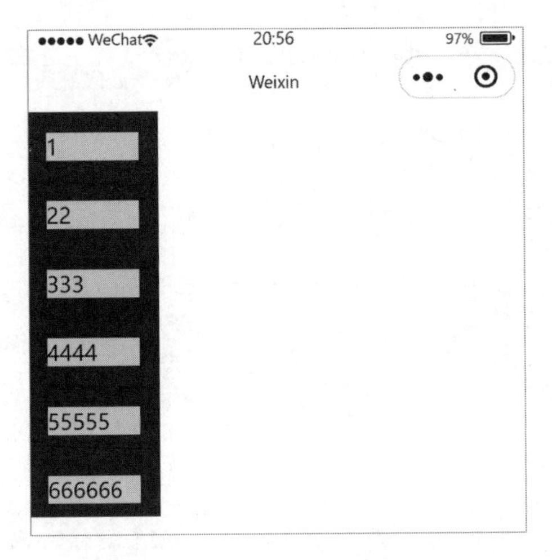

图 2.2.2　Flex 纵向布局

读者可以自行将 flex-direction 属性设置成 row-reverse 和 column-reverse 两个值进行测试，设置后，Flex 项目会改变顺序显示。

为了节省篇幅，后面均以横向布局为例，纵向布局与横向布局类似。

2．测试 flex-wrap 属性

打开 index.wxss 文件，将代码改成：

```
1    .flex-container {
2      display: flex;
3      background: darkblue;
4      height: 200rpx;
5    }
6
7    .flex-container>view {
8      background: lightgray;
9      text-align: center;
10     margin: 50rpx;
11   }
```

为了测试是否换行，第 10 行将外边距设置得比较大。编译后，显示的页面如图 2.2.3 所示。

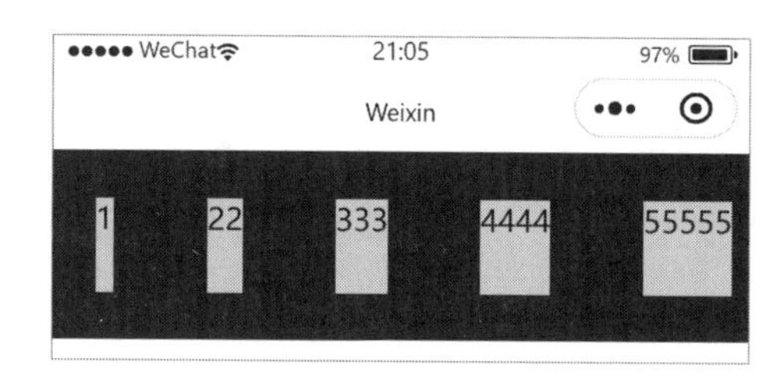

图 2.2.3　Flex 不换行

因为没有换行，最后一项因超出屏幕而无法显示。在.flex-container 样式中增加一行“flex-wrap: wrap;”：

```
1    .flex-container {
2      display: flex;
3      background: darkblue;
4      height: 200rpx;
5      flex-wrap: wrap;
6    }
```

编译后，显示的页面如图 2.2.4 所示。

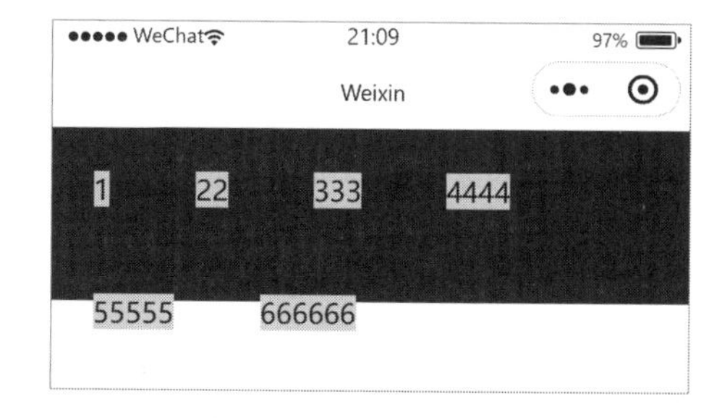

图 2.2.4　Flex 换行

可见，Flex 项目自动换行显示了。读者可以自行测试 flex-wrap 的另一个值 wrap-reverse，以及 flex-flow 属性。

3．测试 justify-content 属性

justify-content 属性是主轴方向上的对齐，默认值是左/上对齐，效果如图 2.2.4 所示。

4．测试 justify-content 属性

打开 index.wxss 文件，将代码改成：

```
1   .flex-container {
2     display: flex;
3     background: darkblue;
4     height: 100rpx;
5     justify-content: flex-end;
6   }
7
8   .flex-container>view {
9     background: lightgray;
10    text-align: center;
11    margin: 20rpx;
12  }
```

第 5 行将 justify-content 属性设置成 flex-end。编译后，显示的页面如图 2.2.5 所示。

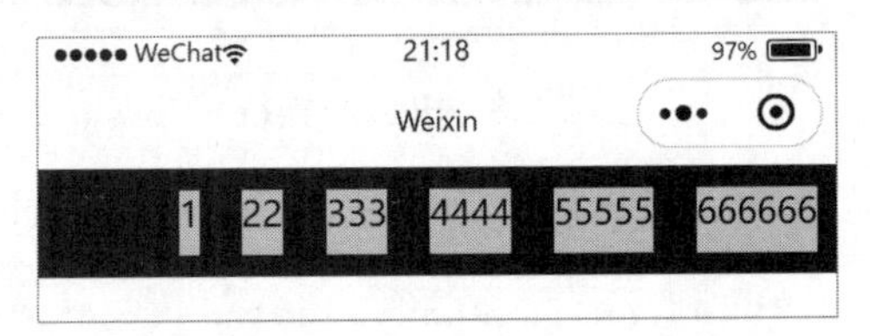

图 2.2.5　Flex 主轴对齐 flex-end

可以看到 Flex 项目右对齐了。其他几个属性值读者可以自行测试。

使用同样的方法可以继续测试 align-items 属性，此处不再赘述。

在大多数情况下，会同时将 justify-content 和 align-items 这两个属性设置成 center，实现“完美对齐”，使界面更美观。

2.2.2　测试 Flex 项目属性

1．测试属性 order

打开 index.wxml 文件，将代码修改为：

```
1   <view class="flex-container">
2     <view style="order: 3;">1</view>
3     <view style="order: 1;">22</view>
```

```
4        <view style="order: 5;">333</view>
5        <view style="order: 6;">4444</view>
6        <view style="order: 2;">55555</view>
7        <view style="order: 4;">666666</view>
8      </view>
```

编译后，显示的页面如图 2.2.6 所示。

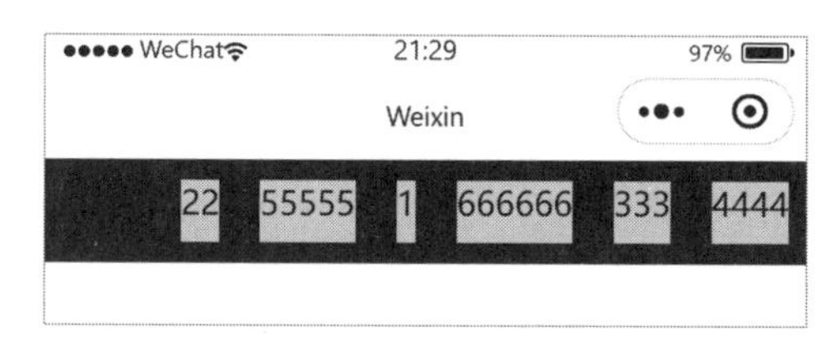

图 2.2.6 order 属性

可以看到 Flex 项目按照 order 属性的顺序显示。

2. 测试属性 flex

测试 Flex 项目最常用的属性是 flex，flex-grow、flex-shrink 和 flex-basis 属性很少单独使用，请读者自行测试。

删除在 index.wxml 文件中添加的 order 属性代码。打开 index.wxss 文件，将代码改为

```
1    .flex-container {
2      display: flex;
3      background: darkblue;
4      height: 100rpx;
5      justify-content: flex-end;
6    }
7
8    .flex-container>view {
9      background: lightgray;
10     text-align: center;
11     margin: 20rpx;
12     flex: 1;
13   }
```

第 12 行添加了 flex 样式。编译后，显示的页面如图 2.2.7 所示。

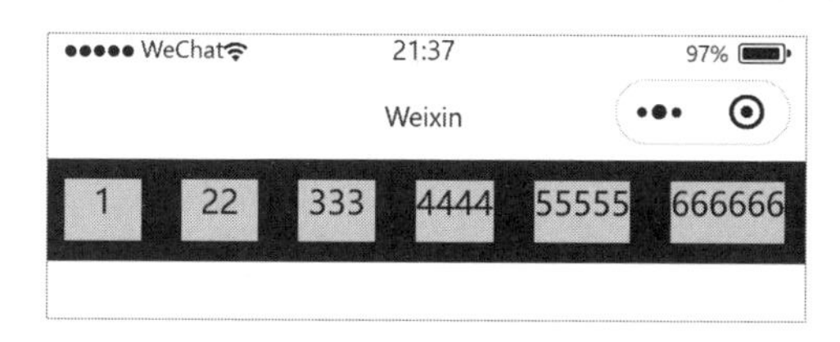

图 2.2.7 flex 属性（1）

可以看出，前 4 个项目的宽度一致，后面两个项目还是原来的宽度，证明 flex 属性影

响的是剩余空间的分配。

打开 index.wxml 文件，将第 2 个 Flex 项目的 flex 改为 2，见下列代码的第 3 行：

```
1    <view class="flex-container">
2      <view>1</view>
3      <view style="flex: 2;">22</view>
4      <view>333</view>
5      <view>4444</view>
6      <view>55555</view>
7      <view>666666</view>
8    </view>
```

编译后，显示的页面如图 2.2.8 所示。

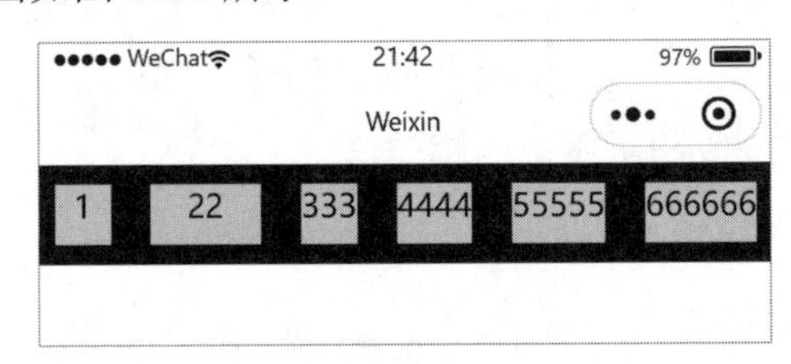

图 2.2.8　flex 属性（2）

可以看出，后 3 个项目保留原来的宽度，第 1 个和第 3 个项目的宽度相同，第 2 个项目的宽度是第 1 个项目宽度的 2 倍。

通过测试，请认真理解 flex 属性的含义：使 Flex 项目平均分配或按比例分配剩余空间。

属性 align-self 留给读者自行测试。

任务三　使用条件渲染和列表渲染

2.3.1　条件渲染

条件渲染是在 WXML 文件中使用 wx:if、wx:elif、wx:else 等关键字，判断是否要渲染某些代码块。

打开微信开发者工具，用测试账号新建一个微信小程序，不使用模板。打开 index.wxml 文件，删除原来的代码，输入以下代码：

```
1    <view class="container">
2      <input type="number" bindchange="change" />
3      <view wx:if="{{score >= 60}}">及格</view>
4      <view wx:else>不及格</view>
5    </view>
```

第 3 行和第 4 行使用了列表渲染。当分数大于或等于 60 时，渲染第 3 行的 view，不渲染第 4 行的 view；否则不渲染第 3 行的 view，渲染第 4 行的 view。第 2 行中的 input 组件的用法在后面的项目中会详细介绍，bindchange 指给 input 组件绑定改变的事件，当 input 中输入的值改变时触发。

打开 index.wxss 文件，删除原来的代码，输入以下代码：

```
1    input {
2      border-bottom: 2rpx #666 solid;
3    }
```

打开 index.js 文件，删除原来的代码，输入以下代码，或者把光标定位于 Page({})的大括号中间，按回车键，再补充代码。

```
1    Page({
2      data: {
3        score: 0
4      },
5      change: function (e) {
6        this.setData({ score: e.detail.value })
7      }
8    })
```

第 5 行中的 change 方法用来处理 input 组件的改变事件，第 6 行中的 e.detail.value 是用户在 input 组件中输入的值。

编译后，分别在 input 输入框中输入 59 和 60 并按回车键，显示的页面如图 2.3.1 所示。

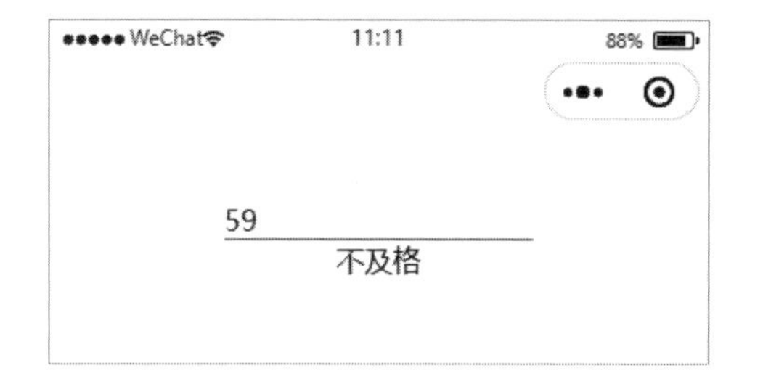

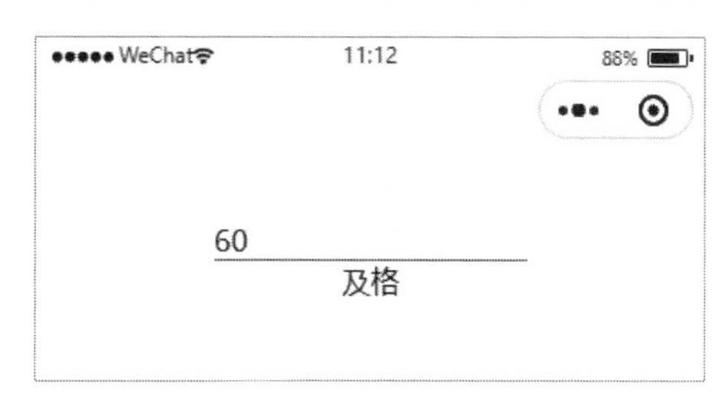

图 2.3.1 条件渲染

还可以继续对分数进行分级，打开 index.wxml 文件，将代码修改为：

```
1    <view class="container">
2      <input type="number" bindchange="change" />
3      <view wx:if="{{score >= 90}}">优秀 YYDS</view>
4      <view wx:elif="{{score >= 70}}">中等</view>
5      <view wx:elif="{{score >= 60}}">及格</view>
6      <view wx:else>不及格</view>
7    </view>
```

读者可以自行输入不同的分数进行测试。

使用条件渲染时，wx:if 关键字是必需的，其他几个关键字可选。

如果多个组件同时使用相同的条件渲染，为多个组件逐个添加 wx:if 关键字会造成重复代码，这种情况下可以使用<block>，如下列代码所示：

```
1    <block wx:if="{{条件}}">
2      <view> view1 </view>
3      <view> view2 </view>
4      <view> view3 </view>
5    </block>
```

当条件满足时，第 2～4 行中的 view 会同时渲染，否则都不渲染，不需要为多个组件逐个添加相同的 wx:if 关键字。

<block>不是一个组件，不会渲染任何内容，只用于渲染控制，在列表渲染中也可以使用。

2.3.2 列表渲染

列表渲染使用 wx:for 作为关键字，绑定一个数组，即可使用数组中的各项数据重复渲染该组件。

打开微信开发者工具，用测试账号新建一个微信小程序，不使用模板。打开 index.wxml 文件，删除原来的代码，输入以下代码：

```
1    <view class="container">
2      <view wx:for="{{list}}" wx:key="name">
3        {{index+1}}: {{item.name}} {{item.score}}
4      </view>
5    </view>
```

第 2～4 行使用了列表渲染，绑定的数组是 list，在 index.js 文件中定义。打开 index.js 文件，输入以下代码：

```
1    Page({
2      data: {
3        list: [
4          { name: '张三', score: 83 },
5          { name: '李四', score: 90 },
6          { name: '王五', score: 76 },
7          { name: '赵六', score: 65 },
8        ]
9      }
10   })
```

第 3～8 行定义了一个数组 list，其中有 4 名学生的姓名和分数。

编译后，显示的页面如图 2.3.2 所示。

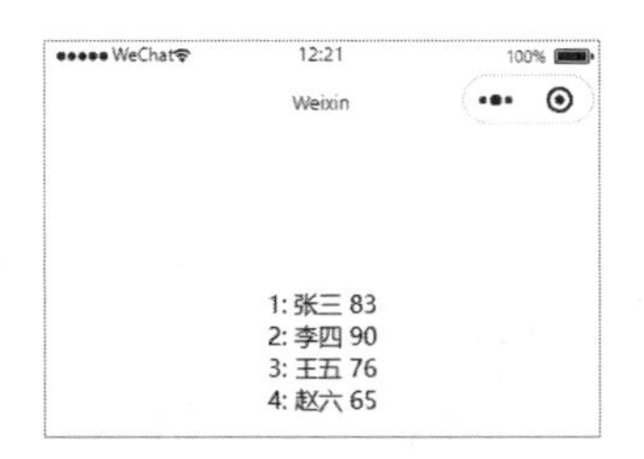

图 2.3.2　列表渲染

数组当前项的下标变量名默认为 index，数组当前项的变量名默认为 item。使用 wx:for-item 可以指定数组当前项的变量名，使用 wx:for-index 可以指定数组当前项的下标的变量名：

```
<view wx:for="{{array}}" wx:for-index="idx" wx:for-item="itemName">
  {{idx}}: {{itemName.message}}
</view>
```

wx:for 关键字还可以嵌套使用，打开 index.wxml 文件，在后面补充以下代码，可以生成九九乘法表：

```
1    <view wx:for="{{[1, 2, 3, 4, 5, 6, 7, 8, 9]}}" wx:for-item="i">
2      <view wx:for="{{[1, 2, 3, 4, 5, 6, 7, 8, 9]}}" wx:for-item="j">
3        <view wx:if="{{i <= j}}">
4          {{i}} * {{j}} = {{i * j}}
5        </view>
6      </view>
7    </view>
```

列表渲染嵌套必须指定不同的数组当前项的变量名，否则会重名。第 1 行和第 2 行分别将当前项的变量名改成了 i 和 j，第 3 行使用了条件渲染。

编译后，显示的页面如图 2.3.3 所示。

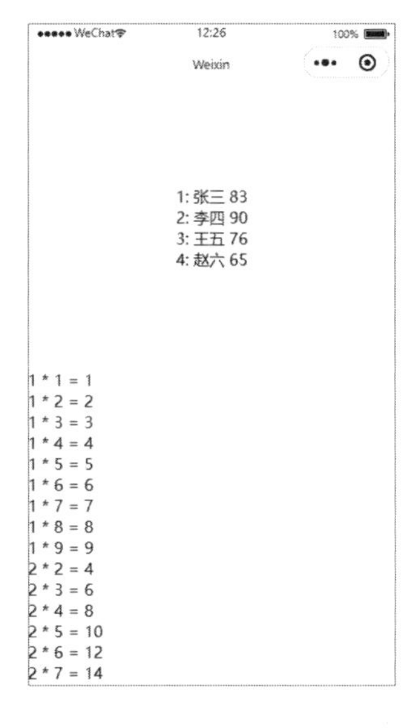

图 2.3.3　列表渲染嵌套

打开调试窗口的 Console 面板，如图 2.3.4 所示，会看到警告信息：为提高性能，可以为 wx:for 关键字提供 wx:key 属性。

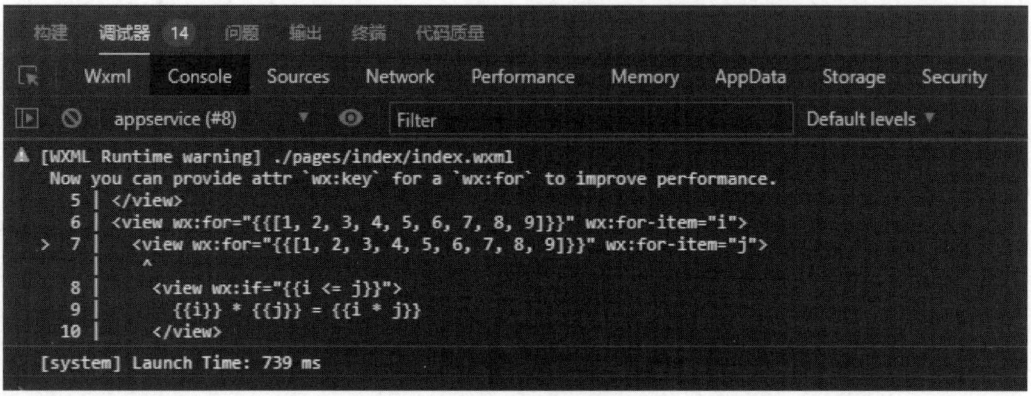

```
[WXML Runtime warning] ./pages/index/index.wxml
 Now you can provide attr `wx:key` for a `wx:for` to improve performance.
   5 | </view>
   6 | <view wx:for="{{[1, 2, 3, 4, 5, 6, 7, 8, 9]}}" wx:for-item="i">
 > 7 |   <view wx:for="{{[1, 2, 3, 4, 5, 6, 7, 8, 9]}}" wx:for-item="j">
     |   ^
   8 |     <view wx:if="{{i <= j}}">
   9 |       {{i}} * {{j}} = {{i * j}}
  10 |     </view>
[system] Launch Time: 739 ms
```

图 2.3.4　Console 面板

index.wxml 文件的第 2 行提供了 wx:key 属性，值是数组元素的 name 属性，因此前面没有此类警告信息。

wx:key 属性用来指定列表中项目的唯一标识符，可以指定数组元素中某个具有唯一值且不变的属性，如前面指定的 name 属性；也可以指定为保留字“*this”，代表在列表渲染中的数组元素本身，这种表示需要数组元素本身是一个唯一的字符串或数字。

<block>同样可以用于列表渲染，如下列代码所示：

```
<block wx:for="{{[1, 2, 3]}}">
  <view> {{index}}: </view>
  <view> {{item}} </view>
</block>
```

项目小结

本项目用一个比较简单的答题小程序来讲解开发微信小程序的一般流程，通过本案例可以了解微信小程序的编程特点，了解页面布局代码、页面样式代码和逻辑代码的区别，以及它们的作用。本项目还讲解了在微信小程序中得到大量应用的 Flex 布局的使用方法，以及在布局代码中经常使用的条件渲染和列表渲染。通过本项目的学习，学生基本可以入门。

习题

一、判断题

1．Flex 布局默认的布局方向是纵向。（　　）

2．使用条件渲染时，关键字 wx:else 是必需的。（　　）

3．使用列表渲染时，关键字 wx:for 必须绑定一个数组。（　　）

二、选择题

1．若将 Flex 布局方向设置为纵向，需要设置 flex-direction 为（　　）。

A．row

B．row-reverse

C．column

D．column-reverse

2．若将 Flex 项目设置为主轴居中对齐，需要将（　　）属性设置为 center。

A．justify-content

B．align-items

C．flex

D．flex-flow

3．在列表渲染中，（　　）用于设置当前项的下标变量名。

A．wx:for

B．wx:key

C．wx:for-index

D．wx:for-item

三、填空题

1．微信小程序生命周期一般包括创建、装载、显示、隐藏、唤醒、停止、____、销毁等。

2．加载页面时，触发页面的__________方法，一个页面只调用一次。

3．初次渲染页面时，触发页面的__________方法渲染页面元素和样式，一个页面只调用一次。

四、编程题

参照答题小程序，扩展功能，实现选择题的答题（可将判断题作为选择题的一种）。

项目三　天气预报

任务一　使用 Picker 组件

3.1.1　Picker 组件简介

Picker 组件（滚动选择器）属于表单大类的组件，主要用于准确地接受用户的滚动选择并输入结果。根据选择内容的不同，Picker 组件分为普通选择器、时间选择器、日期选择器、多列选择器和省市区选择器。picker 标签中的 mode 属性决定组件的类型，mode 属性与选择器类型关系如表 3.1.1 所示。

表 3.1.1　mode 属性与选择器类型关系

mode 属性	选择器类型
mode=selector	普通选择器，默认类型
mode=multiSelector	多列选择器
mode=date	日期选择器
mode=time	时间选择器
mode=region	省市区选择器

3.1.2　Picker 组件的使用

1. 普通选择器

当 picker 标签中的 mode 属性为 selector，或者不指定 mode 属性时，Picker 组件为普通选择器。range 属性指定被选择项的数据源，bindchange 绑定选择项发生改变时对应的 js 处理函数。普通选择器的运行效果如图 3.1.1 所示。

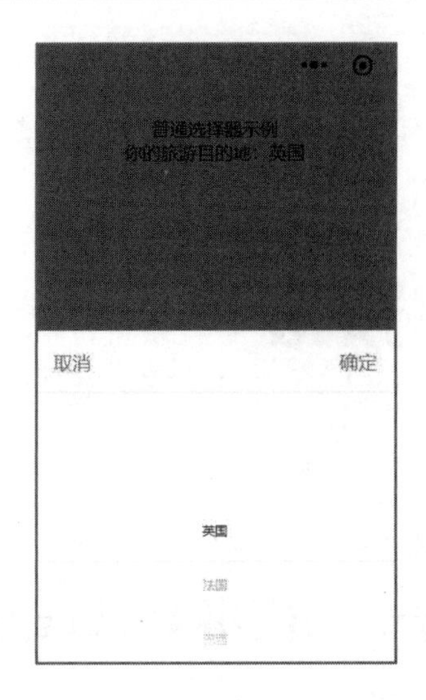

图 3.1.1　普通选择器的运行效果

要实现这个效果，可以打开微信开发者工具，使用测试账号新建一个微信小程序，不使用模板。打开 index.wxml 文件，删除原来的代码，输入以下代码：

```
1   <view class="container">
2   <view >普通选择器示例</view>
3     <picker bindchange="bindPickerChange" value="{{index}}" range="
    {{array}}">
4       <view class="picker">
5         你的旅游目的地：{{array[index]}}
6       </view>
7     </picker>
8   </view>
```

在上述代码中，picker 标签没有指定 mode 属性，因此是默认的普通选择器；第 3 行中的 range 属性指定选择项的数据源为数组 array。

打开 index.js 文件，删除原来的代码，输入以下代码：

```
1   Page({
2     data: {
3       array: ['英国', '法国', '巴西', '德国'],
4       index: 0
5     },
6     //选择项发生变化时对应的处理函数
7     bindPickerChange(e){
8       this.setData({
9       //e.detail.value 是被选择项的索引号
10        index:e.detail.value
11      })
12    }
13  })
```

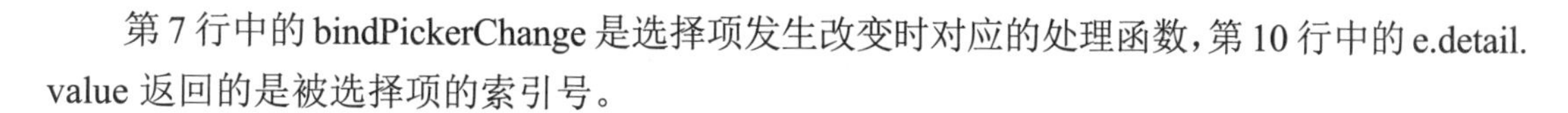

第 7 行中的 bindPickerChange 是选择项发生改变时对应的处理函数，第 10 行中的 e.detail.value 返回的是被选择项的索引号。

2．多列选择器

当 picker 标签中的 mode 属性为 multiSelector 的时候，Picker 组件为多列选择器。range 属性用于指定多列数据的多维数组数据源，bindchange 绑定选择项发生改变时对应的 js 处理函数，bindcolumnchange 绑定多列中某个选择项发生改变的那一列对应的 js 处理函数。多列选择器的运行效果如图 3.1.2 和图 3.1.3 所示。

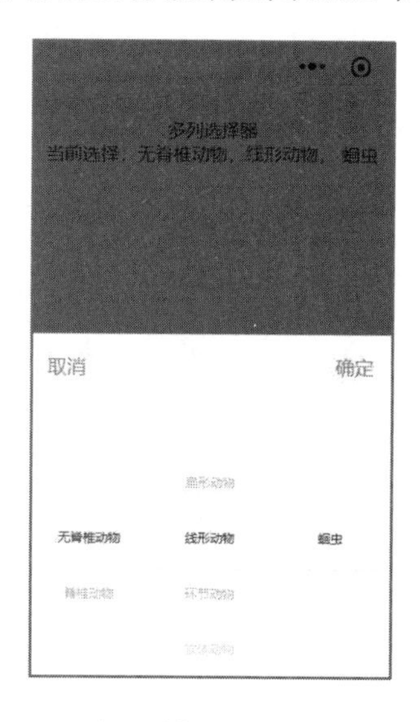

图 3.1.2　多列选择器的运行效果（1）

图 3.1.3　多列选择器的运行效果（2）

要实现这个效果，可以打开微信开发者工具，使用测试账号新建一个微信小程序，不使用模板。打开 index.wxml 文件，删除原来的代码，输入以下代码：

```
1    <view class="container">
2      <view>多列选择器</view>
       <picker mode="multiSelector" bindchange="bindMultiPickerChange"
3  bindcolumnchange="bindMultiPickerColumnChange" value="{{multiIndex}}"
   range="{{multiArray}}">
         <view class="picker"> 当前选择：{{multiArray[0][multiIndex[0]]}},
4  {{multiArray[1][multiIndex[1]]}},
5      {{multiArray[2][multiIndex[2]]}} </view>
6      </picker>
7    </view>
```

第 3 行中的 mode="multiSelector"说明这是一个多列选择器，bindchange 绑定选择项发生改变时对应的处理函数是"bindMultiPickerChange"，bindcolumnchange 绑定选择列发生改变时对应的处理函数是"bindMultiPickerColumnChange"。

打开 index.js 文件，删除原来的代码，输入以下代码：

```
1    Page({
2      data: {
3        multiArray: [['无脊椎动物', '脊椎动物'], ['扁形动物', '线形动物', '环
     节动物', '软体动物', '节肢动物'], ['猪肉绦虫', '血吸虫']],
4        multiIndex: [0, 0, 0]
5      },
6      bindMultiPickerChange: function (e) {
7        console.log('picker发送选择改变，携带值为', e.detail.value)
8        this.setData({ multiIndex: e.detail.value  })
9      },
10     bindMultiPickerColumnChange: function (e) {
11       console.log('修改的列为', e.detail.column, '，值为', e.detail.value);
12       var data = {
13         multiArray: this.data.multiArray,
14         multiIndex: this.data.multiIndex
15       };
16       data.multiIndex[e.detail.column] = e.detail.value;
17       switch (e.detail.column) {
18         case 0:
19           switch (data.multiIndex[0]) {
20             case 0:
21               data.multiArray[1] = ['扁形动物', '线形动物', '环节动物', '软
     体动物', '节肢动物'];
22               data.multiArray[2] = ['猪肉绦虫', '血吸虫'];
23               break;
24             case 1:
25               data.multiArray[1] = ['鱼', '两栖动物', '爬行动物'];
26               data.multiArray[2] = ['鲫鱼', '带鱼'];
27               break;
28           }
29           data.multiIndex[1] = 0;
30           data.multiIndex[2] = 0;
31           break;
32         case 1:
33           switch (data.multiIndex[0]) {
34             case 0:
35               switch (data.multiIndex[1]) {
36           case 0: data.multiArray[2] = ['猪肉绦虫', '血吸虫']; break;
37                 case 1:data.multiArray[2] = ['蛔虫']; break;
38                 case 2:data.multiArray[2] = ['蚂蚁', '蚂蟥']; break;
39                 case 3: data.multiArray[2] = ['河蚌', '蜗牛', '蛞蝓']; break;
40                 case 4: data.multiArray[2] = ['昆虫', '甲壳动物', '蛛形动
     物', '多足动物']; break;
41               }
42               break;
43               case 1:
```

```
44              switch (data.multiIndex[1]) {
45                case 0: data.multiArray[2] = ['鲫鱼', '带鱼']; break;
46                case 1:data.multiArray[2] = ['青蛙', '娃娃鱼']; break;
47                case 2:data.multiArray[2] = ['蜥蜴', '龟', '壁虎']; break; }
48              break;
49          }
50          data.multiIndex[2] = 0;
51          break; }
52      console.log(data.multiIndex);
53      this.setData(data);
54    }
55  })
```

第 3 行定义了一个二维数组 multiArray，multiArray[0]对应的是第一列选择器的默认数据源，multiArray[1]对应的是第二列选择器的默认数据源，multiArray[2]对应的是第三列选择器的默认数据源；第 4 行中的 multiIndex 是一个数组，其中的元素代表三列选择器当前被选中项的索引号；第 6 行中的 bindMultiPickerChange 是多列选择器选项发生改变时对应的处理函数；第 10 行中的 bindMultiPickerColumnChange 是多列选择器的某列发生改变时对应的处理函数，程序会根据当前选择的列和项重置第二列和第三列选择器的数据源数组。

3. 日期选择器

当 picker 标签中的 mode 属性为 date 的时候，Picker 组件为日期选择器。start 属性指可以选择的开始日期，end 属性为可以选择的截止日期，bindchange 绑定选择项发生改变时对应的 js 处理函数。日期选择器的运行效果如图 3.1.4 所示。

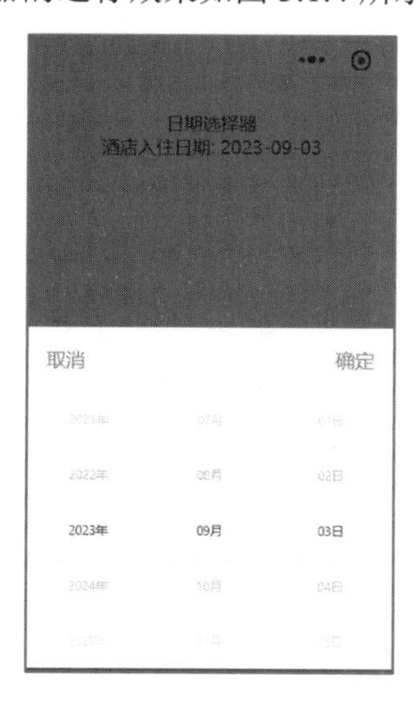

图 3.1.4　日期选择器的运行效果

要实现这个效果，可以打开微信开发者工具，使用测试账号新建一个微信小程序，不使用模板。打开 index.wxml 文件，删除原来的代码，输入以下代码：

```
<view class="container">
  <view >日期选择器</view>
  <picker mode="date" value="{{date}}" start="2023-09-01" end="2023-09-09" bindchange="bindDateChange">
    <view class="picker">    酒店入住日期: {{date}}    </view>
  </picker>
</view>
```

在上述代码中，mode 属性为 date，说明这是一个日期选择器；start 属性的值是可以选择的开始日期，end 属性的值是可以选择的截止日期，value 属性的值是选择器的当前值。

打开 index.js 文件，删除原来的代码，输入以下代码：

```
Page({
  data: {
    date: '2023-09-03'
  },
  //选择的日期发生变化时对应的处理函数
  bindDateChange: function(e) {
    this.setData({
      date: e.detail.value
    })
  }
})
```

第 3 行中的 date 变量的初始值是“2023-09-03”，第 6 行中的 bindDateChange 是选择器的值发生变化时对应的处理函数。

4. 时间选择器

当 picker 标签中的 mode 属性为 time 的时候，Picker 组件为时间选择器。start 属性指可以选择的开始时间(格式为hh:mm，即小时:分钟)，end属性为可以选择的截止时间，bindchange 绑定选择项发生改变时对应的 js 处理函数。时间选择器的运行效果如图 3.1.5 所示。

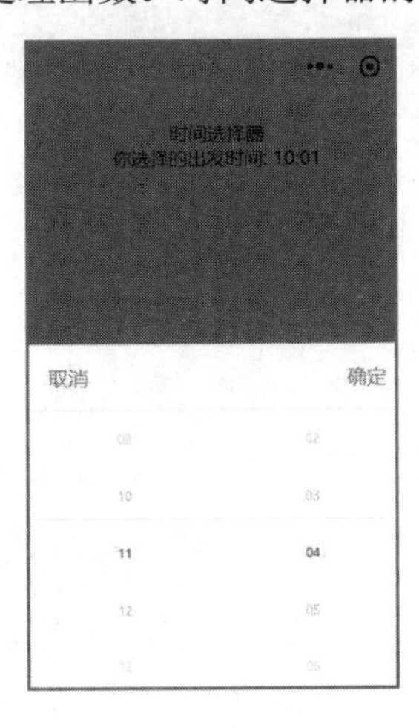

图 3.1.5　时间选择器的运行效果

要实现这个效果，可以打开微信开发者工具，使用测试账号新建一个微信小程序，不使用模板。打开 index.wxml 文件，删除原来的代码，输入以下代码：

```
1    <view class="container">
2      <view >时间选择器</view>
3      <picker mode="time" value="{{time}}" start="09:01" end="11:01" 
     bindchange="bindTimeChange">
4        <view class="picker"> 你选择的出发时间：{{time}}   </view>
5      </picker>
6    </view>
```

第 3 行中的 mode 属性为 time 决定了这是一个时间选择器，start 属性的值是可以选择的开始时间，end 属性的值是可以选择的结束时间，value 属性的值是当前选择的值。

打开 index.js 文件，删除原来的代码，输入以下代码：

```
1    Page({
2      data: {
3        time: '10:01'
4      },
5    //选择的时间发生改变时对应的处理函数
6      bindTimeChange: function(e) {
7        this.setData({
8          time: e.detail.value
9        })
10     }
11   })
```

5. 省市区选择器

当 picker 标签中的 mode 属性为 region 的时候，Picker 组件为省市区选择器。省市区选择器的数据源是自动获取的，无须开发人员指定。value 属性的值为当前选择的值。与其他选择器不同的是，省市区选择器中的 custom-item 可为每列的顶部添加一个自定义项，bindchange 绑定选择项发生改变时对应的 js 处理函数。省市区选择器的运行效果如图 3.1.6 所示。

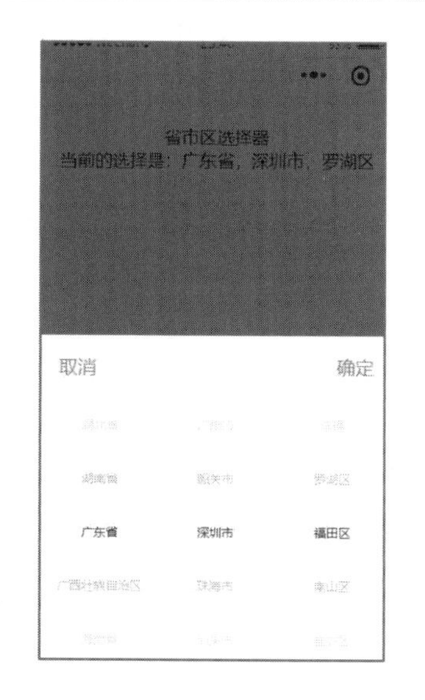

图 3.1.6　省市区选择器的运行效果

要实现这个效果，可以打开微信开发者工具，使用测试账号新建一个微信小程序，不使用模板。打开 index.wxml 文件，删除原来的代码，输入以下代码：

```
1    <view class="container">
2      <view >省市区选择器</view>
```

```
3      <picker mode="region" bindchange="bindRegionChange" value=
   "{{region}}" custom-item="{{customItem}}">
4        <view class="picker">
5          当前的选择是：{{region[0]}}，{{region[1]}}，{{region[2]}}
6        </view>
7      </picker>
8    </view>
```

第 3 行中的 mode="region"表示这是一个省市区三级联动选择器；第 5 行中的 region[0]、region[1]、region[2]分别是省、市、区的数据源数组。第 3 行中的 bindchange="bindRegionChange"把选择器的选择变化事件和 js 代码中的"bindRegionChange"函数进行了绑定。

打开 index.js 文件，删除原来的代码，输入以下代码：

```
1    Page({
2      data: {
3        region: ['广东省', '深圳市', '罗湖区'],
4        customItem: '选择'
5      },
6      bindRegionChange: function (e) {
7        console.log('picker当前值是:', e.detail.value)
8        this.setData({
9          region: e.detail.value
10       })
11     }
12   })
```

第 3 行中的 region 是保存当前选择的值的数组。第 6 行中的 bindRegionChange 绑定选择项发生变化的事件的对应处理函数。

任务二　使用 wx.request 发起网络请求

3.2.1　请求服务器数据 API

为了方便开发者开发微信小程序，腾讯把很多常用功能都封装成 API 接口，以便开发者调用，如请求服务器数据 API、文件上传下载 API、图片处理 API、文件操作 API、设备应用 API 等。wx.request 是用于请求服务器数据的 API，调用这个接口会发起一个 HTTPS 请求。在小程序请求服务器数据之前，要在小程序的后台做好以下设置。

小程序管理员登录微信公众平台管理后台后，在管理界面中按照“开发-开发管理-开发设置-服务器域名设置”的顺序进入服务器域名配置项，配置界面如图 3.2.1 所示。

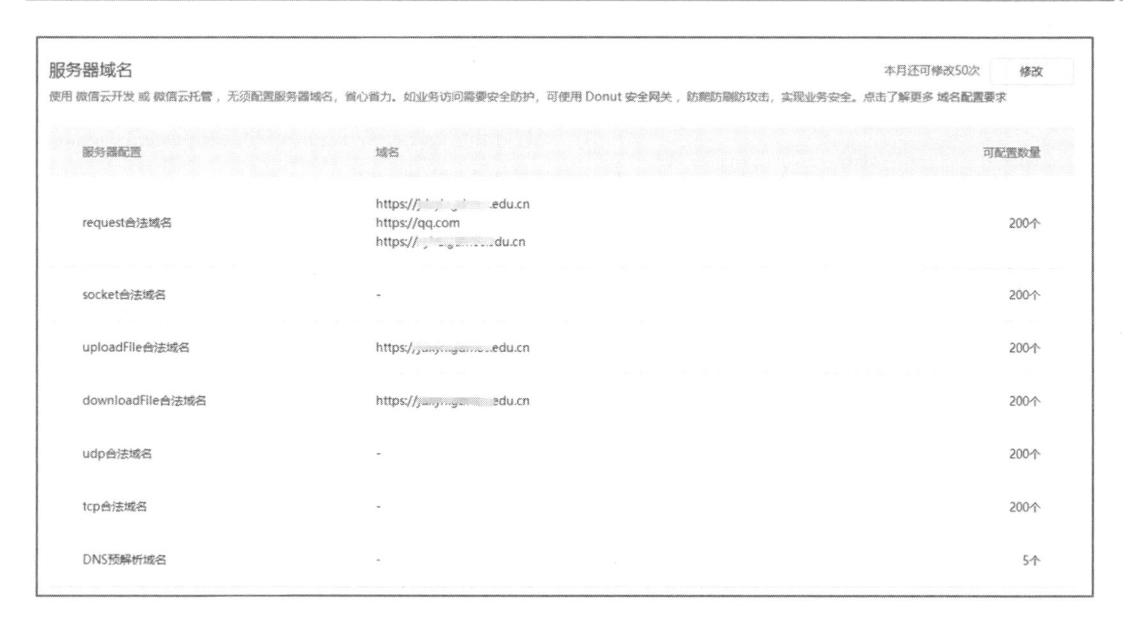

图 3.2.1 小程序服务器域名配置界面

服务器域名配置的注意事项如下。

（1）域名只支持 HTTPS（wx.request、wx.uploadFile、wx.downloadFile 接口使用）和 WSS（wx.connectSocket）。为了支持 HTTPS，被访问的数据服务器应安装 HTTPS 证书和 SSL 证书。

（2）域名不能使用 IP 地址（小程序的局域网 IP 除外）或 localhost。

（3）域名必须经过 ICP 备案。

（4）出于安全考虑，api.weixin.qq.com 不能被配置为服务器域名，相关 API 也不能在小程序内调用。开发者应将 AppSecret 保存到后台服务器中，通过服务器使用 getAccessToken 接口获取 access_token，并调用相关 API。

3.2.2 wx.request 请求参数

调用 wx.request 接口时需要先设置请求参数和回调函数，相关参数如表 3.2.1 所示，url、data、header 和 method 是主要的请求参数。其中，url 是要请求的网络服务的地址；data 封装要请求的数据的条件，如查询商品编号、查询学号和课程编号等。success、fail 和 complete 分别是接口调用成功、失败和结束时的 js 回调处理函数。

表 3.2.1 wx.request 接口参数

属性	类型	是否必填	说明
url	string	是	要请求的网络服务的地址
data	string/object/ArrayBuffer	否	要请求的数据的条件

续表

属性	类型	是否必填	说明
header	object	否	设置请求的 header。在 header 中不能设置 Referer。content-type 默认为 application/json
method	string	否	默认为 GET。有效值为 OPTIONS、GET、HEAD、POST、PUT、DELETE、TRACE 和 CONNECT
success	function	否	接口调用成功的回调函数，res = {data: '开发者服务器返回的内容'}
fail	function	否	接口调用失败的回调函数
complete	function	否	接口调用结束的回调函数（调用成功或调用失败都会执行）

3.2.3 wx.quest 请求示例

要调用网络请求的接口，首先要在小程序管理后台做好设置，然后在调用接口时做好 url、data 等请求参数及 success、fail 等回调函数的设置。下面来看一个简单例子的 js 代码：

```
1  Page({
2    onLoad: function () {
3      wx.request({
4        //url 为要请求的网络服务的地址
5        url: 'https://www.edugaaa.com/xcx/index.php/Home/Getda/test22',
6
7        data: {
8          //name 为封装在接口中的请求参数
9          name: 'jack'
10       },
11       //POST 为网络请求的方式，也可以使用 GET 等方式
12       method: 'POST',
13       //success 是请求成功后的回调函数
14       success: function (res) {
15         console.log(res);
16       },
17       //fail 是请求失败后的回调函数
18       fail: function (ff) {
19         console.log(ff.errMsg);
20       },
21       //complete 是请求结束后的回调函数
22       complete: function (cc) {
23         console.log('网络请求结束');
24       }
25     });
26   }
27 })
```

第 3 行中的 wx.request 是发送网络请求时要调用的 API，第 5 行中的 url 是要请求的网络服务的地址，第 7～10 行的 data 中有一个值为“jack”、名为 name 的输入参数。本段代码还定义了接口调用成功、失败和完成时要调用的回调函数。

任务三　制作天气预报小程序

3.3.1　项目介绍

天气预报小程序使用和风天气开发服务，可以查询全国各地实时天气、24 小时天气预报和 7 天天气预报。先使用省市区选择器选择省市区，然后显示该区域的实时天气、24 小时天气预报和 7 天天气预报。24 小时天气预报和 7 天天气预报屏幕显示不下，可以通过水平移动滚动条查看。天气预报小程序如图 3.3.1 所示。

图 3.3.1　天气预报小程序

3.3.2　和风天气开发服务简介

和风天气开发服务为开发者和企业提供了强大、丰富的天气数据服务，可以在应用中展示天气。

要使用和风天气开发服务，需要先注册。进入和风天气开发服务的首页，单击“免费注册”按钮，如图 3.3.2 所示。

图 3.3.2 和风天气开发服务

先按照提示免费注册账号，再创建项目和 key，注意选择 key 平台时必须选择 Web API。注册完成后，保存注册的 key，在小程序的 js 代码中需要使用。

进入微信开发平台，并登录自己的小程序账号，先进入“开发|开发管理”界面，再进行开发设置，将和风天气开发服务的首页网址加入 request 合法域名。

3.3.3 创建项目

打开微信开发者工具，新建项目，名称为 weather，使用自己注册的 AppID，不使用模板。打开 app.json 文件，修改“navigationBarTitleText”为“天气预报”。

先复制资源文件夹中 weather 下面的 128 文件夹到新建的 weather 微信小程序文件夹，该文件夹与 pages 文件夹并列。128 文件夹中有一批天气图标。

再复制资源文件夹中 weather 下面的 location.js 到 index 文件夹，与 index.js 并列。

3.3.4 查看实时天气

扫一扫

微课：天气预报代码 1

打开 index.wxml 文件，删除原来的代码，输入以下代码：

```
1   <view class="header">
2     <view class="top">
3       <view class="search">
4         <picker mode="region" bindchange="bindRegionChange" value=
"{{region}}" custom-item="{{customItem}}">
5           <view>
6             {{region[0]}}, {{region[1]}}, {{region[2]}}
```

```
7            </view>
8          </picker>
9        </view>
10     </view>
11     <view class="center">
12       <view class="tmp"> {{now.temp}}° </view>
13       <view class="cond-image">
14         <image mode="widthFix" src="/128/{{now.icon}}.png" />
15         <view>{{now.text}}</view>
16       </view>
17     </view>
18     <view class="bottom">
19       <view>{{now.windDir}} {{now.windScale}}级</view>
20       <view>湿度 {{now.humidity}}%</view>
21       <view>气压 {{now.pressure}}Pa</view>
22     </view>
23   </view>
```

now.icon 是得到的天气代码，第 14 行 image 组件的 src 属性使用字符串拼接的方式找到文件夹 128 下面对应的天气图标并显示。24 小时天气预报和 7 天天气预报均使用类似的方法显示天气图标。

打开 index.wxss 文件，输入以下代码：

```
1    page {
2      background-color: #f6f6f6;
3    }
4
5    .header {
6      background-color: #64c8fa;
7      height: 500rpx;
8      padding-top: 32rpx;
9      text-align: center;
10   }
11
12   .top {
13     display: flex;
14     justify-content: space-between;
15     align-content: center;
16     align-items: center;
17   }
18
19   .search {
20     margin: 0 32rpx;
21     border-radius: 8rpx;
22     background-color: rgba(0, 0, 0, 0.3);
23     height: 80rpx;
```

```
24      }
25
26      .search>picker {
27        width: 630rpx;
28        padding: 18rpx 32rpx;
29        color: white;
30      }
31
32      .center {
33        display: flex;
34        justify-content: space-between;
35        align-content: center;
36        align-items: center;
37      }
38
39      .tmp {
40        margin-left: 18rpx;
41        display: inline-block;
42        font-size: 180rpx;
43        color: white;
44      }
45
46      .cond-image {
47        display: flex;
48        flex-direction: column;
49        align-content: center;
50        width: 200rpx;
51        margin-right: 32rpx;
52        margin-top: 12rpx;
53      }
54
55      .cond-image>image {
56        width: 200rpx;
57      }
58      .cond-image>view {
59        color: white;
60      }
61
62      .bottom {
63        display: flex;
64        justify-content: space-between;
65        align-content: center;
66        align-items: center;
67      }
68
69      .bottom>view {
```

```
70      color: white;
71      margin: 32rpx;
72    }
```

打开 index.js 文件，在 Page({})的前面输入以下代码：

```
1   const key = ''
2   const location = require('./location.js')
```

在第 1 行填入注册和风天气开发服务时得到的 key。第 2 行代码用于引入 location.js 模块。在 Page({})的里面输入以下代码：

```
1      data: {
2        region: ['广东省', '广州市', '海珠区'],
3        customItem: '全部',
4        now: {},
5      },
6
7      cityId: '',
8      onLoad: function(e) {
9        this.cityId = '101280101' //广州市海珠区
10       this.weather()
11     },
12
13     bindRegionChange: function (e) {
14       console.log('picker 发送选择改变，携带值为', e)
15       let arr = location.getLocation_ID(e.detail.code[e.detail.code.
   length - 1])
16       if(arr.length == 0) {
17         wx.showToast({
18           title: '查询失败',
19           duration: 1500
20         })
21         return
22       }
23       this.cityId = arr[0]
24       console.log(this.cityId)
25       this.weather()
26       this.setData({
27         region: e.detail.value
28       })
29     },
30
31     weather: function() {
32       this.weather_now()
33     },
34
35     weather_now: function() {
```

```
36        wx.showLoading({
37          title: '加载中',
38        })
39        wx.request({
40          url: 'https://devapi.qweather.com/v7/weather/now?key=' + key +
   '&location=' + this.cityId,
41          success: res => {
42            console.log(res)
43            if (!res.data.now) {
44              console.log('获取天气接口失败')
45              wx.hideLoading()
46              return
47            }
48            this.setData({
49              now: res.data.now,
50            })
51            wx.hideLoading()
52          },
53          fail: err => {
54            console.log(err)
55            wx.hideLoading()
56          }
57        })
58      },
```

第 15 行中的 e.detail.code 是一个数组，最多包含 3 个元素，分别代表省、市、区对应的行政区划代码。用行政区划代码调用 location.getLocation_ID 查找对应的省、市、区的数据，并返回一个数组，数组的第一个元素就是查询和风天气需要的 Location_ID。

使用和风天气查询实时天气至少需要指定以下两个参数。

（1）key（必选），用户认证 key，就是注册时保存的 key。

（2）location（必选），需要查询地区的 Location_ID 或以英文逗号分隔的经度、纬度坐标（十进制，最多支持小数点后两位）。

查询到的结果保存在第 49 行的 res.data.now 中，将查询结果保存到 now 后直接在页面中显示即可。now 中各属性的含义可以查看和风天气文档。

wx.showLoading 显示 loading 提示框，需主动调用 wx.hideLoading 才能关闭提示框。wx.showLoading 通常用于耗时的操作，耗时操作开始时显示提示信息，耗时操作结束时关闭。

wx.showToast 显示消息提示框，超过指定时间会自动消失。

编译运行后，可以选择省市区，并可以查看实时天气。

3.3.5　查看 24 小时天气预报

扫一扫

微课：天气预报代码 2

打开 index.wxml 文件，在后面输入以下代码：

```
1    <view class="title">24 小时预报</view>
2    <scroll-view scroll-x="true" class="hourly">
3      <view class="h_item" wx:for="{{hourly}}" wx:key="index">
4        <text class="h_time">{{item.time}}</text>
5        <view class="h_img">
6          <image mode="widthFix" src="/128/{{item.icon}}.png" />
7        </view>
8        <text class="h_text">{{item.text}}</text>
9        <text class="h_tmp">{{item.temp}}°</text>
10       <text class="h_wind_dir">{{item.windDir}}</text>
11       <text class="h_wind_sc">{{item.windScale}}级</text>
12     </view>
13   </scroll-view>
```

第 3～12 行使用列表渲染显示 hourly，循环显示未来 24 小时的天气情况。

打开 index.wxss 文件，在后面输入以下代码：

```
1    .title {
2      font-weight: bold;
3      font-size: 42rpx;
4      padding: 18rpx 32rpx;
5    }
6
7    .hourly {
8      width: 718rpx;
9      margin: 0 18rpx;
10     border-radius: 18rpx;
11     box-shadow: 0.1rem 0.1rem 0.5rem rgba(0, 0, 0, 0.15);
12     white-space: nowrap;
13     background-color: white;
14   }
15
16   .h_item {
17     margin: 18rpx 0;
18     display: inline-block;
19     width: 143.5rpx;
20     text-align: center;
21     font-size: 28rpx;
22   }
23
24   .h_img {
25     margin: 32rpx 0;
26   }
```

```
27
28    .h_img>image {
29      width: 100rpx;
30    }
31
32    .h_item>text {
33      display: block;
34    }
35
36    .h_time {
37      color: gray;
38    }
39
40    .h_text {
41      color: cadetblue;
42    }
43
44    .h_wind_dir {
45      margin-top: 12rpx;
46    }
47
48    .h_wind_sc {
49      color: gray;
50    }
51
52    .h_tmp {
53      color: #027aff;
54    }
```

打开 index.js 文件，在 weather 方法中添加 this.weather_24h()，见下面代码的粗斜体部分：

```
weather: function() {
  this.weather_now()
  this.weather_24h()
},
```

在 weather_now 方法的后面添加 weather_24h 方法：

```
1      weather_24h: function() {
2        wx.showLoading({
3          title: '加载中',
4        })
5        wx.request({
6          url: 'https://devapi.qweather.com/v7/weather/24h?key=' + key +
   '&location=' + this.cityId,
7          success: res => {
8            console.log(res)
9            if (!res.data.hourly) {
```

```
10              console.log('获取天气接口失败')
11              wx.hideLoading()
12              return
13            }
14            res.data.hourly.forEach(h => {
15              h.time = h.fxTime.substring(11, 16)
16            });
17            this.setData({
18              hourly: res.data.hourly
19            })
20            wx.hideLoading()
21          },
22          fail: err => {
23            console.log(err)
24            wx.hideLoading()
25          }
26        })
27      },
```

逐小时天气预报数据通过 res.data.hourly 返回，hourly 是一个数组，包含 24 个元素，代表未来 24 小时的天气预报。h.fxTime 是预报时间，如 2021-02-16T16:00+08:00。代码的第 14～16 行对 hourly 数组进行 forEach 循环，为数组的每个元素添加了一个 time 属性，截取 fxTime 字符串的第 11～16 个字符串，即 fxTime 的时分部分显示在页面中。

编译运行后，可以查看未来 24 小时的天气预报。

扫一扫

3.3.6　查看 7 天天气预报

微课：天气预报代码 3

打开 index.wxml 文件，在后面输入以下代码：

```
1      <view class="title">7 天预报</view>
2      <scroll-view scroll-x="true" class="daily">
3        <view class="d_item" wx:for="{{daily}}" wx:key="index">
4          <text class="d_txt">{{item.week}}</text>
5          <text class="d_date">{{item.date}}</text>
6          <text class="d_wind_dir">{{item.windDirDay}}</text>
7          <text class="d_wind_sc">{{item.windScaleDay}}级</text>
8          <text class="d_text">{{item.textDay}}</text>
9          <view class="d_img">
10           <image mode="widthFix" src="/128/{{item.iconDay}}.png" />
11         </view>
12         <text class="d_tmp">{{item.tempMin}}°~{{item.tempMax}}°</text>
13         <view class="d_img">
14           <image mode="widthFix" src="/128/{{item.iconNight}}.png" />
15         </view>
16         <text class="d_text">{{item.textNight}}</text>
```

```
17        <text class="d_wind_dir">{{item.windDirNight}}</text>
18        <text class="d_wind_sc">{{item.windScaleNight}}级</text>
19      </view>
20    </scroll-view>
21
22    <view class="footer">天气数据来自和风天气</view>
```

第3～19行使用列表渲染显示daily，循环显示未来7天的天气情况。

打开index.wxss文件，在后面输入以下代码：

```
1     .daily {
2       width: 718rpx;
3       white-space: nowrap;
4       margin: 0 18rpx;
5       background-color: white;
6       border-radius: 18rpx;
7       box-shadow: 0.1rem 0.1rem 0.5rem rgba(0, 0, 0, 0.15);
8     }
9
10    .d_item {
11      margin: 18rpx 0;
12      display: inline-block;
13      width: 179.5rpx;
14      text-align: center;
15      font-size: 28rpx;
16    }
17
18    .d_item text {
19      display: block;
20    }
21
22    .d_date {
23      color: gray;
24    }
25
26    .d_wind_dir {
27      margin: 16rpx 0;
28    }
29
30    .d_wind_sc {
31      color: gray;
32    }
33
34    .d_text {
35      color: cadetblue;
36    }
37
```

```
38      .d_img {
39        margin: 16rpx 0;
40      }
41
42      .d_img>image {
43        width: 100rpx;
44      }
45
46      .d_tmp {
47        color: #027aff;
48      }
49
50      .footer {
51        font-size: 28rpx;
52        color: gray;
53        text-align: center;
54        margin-top: 50rpx;
55        margin-bottom: 18rpx;
56      }
```

打开 index.js 文件，在 weather 方法中添加 this.weather_7d()，见下面代码的粗斜体部分：

```
weather: function() {
  this.weather_now()
  this.weather_24h()
  this.weather_7d()
},
```

在 Page({})的前面补充以下代码：

```
const weekArray = new Array("周日", "周一", "周二", "周三", "周四", "周五", "周六")
```

在 weather_24h 方法的后面添加 weather_7d 方法：

```
1        weather_7d: function() {
2          wx.showLoading({
3            title: '加载中',
4          })
5          wx.request({
6            url: 'https://devapi.qweather.com/v7/weather/7d?key=' + key + '&location=' + this.cityId,
7            success: res => {
8              console.log(res)
9              if (!res.data.daily) {
10               console.log('获取天气接口失败')
11               wx.hideLoading()
12               return
13             }
```

```
14            res.data.daily.forEach(d => {
15              let date = new Date(d.fxDate)
16              d.week = weekArray[date.getDay()]//getDay 方法返回 0～6，代表星期
17              d.date = d.fxDate.substring(5)
18            });
19            res.data.daily[0].week = '今日'
20            this.setData({
21              daily: res.data.daily
22            })
23            wx.hideLoading()
24          },
25          fail: err => {
26            console.log(err)
27            wx.hideLoading()
28          }
29        })
30      }
```

每天的天气预报数据通过 res.data.daily 返回，daily 是一个数组，包含 7 个元素，代表未来 7 天的天气预报。d.fxDate 是预报日期。代码的第 14～18 行对 daily 数组进行 forEach 循环，为数组的每个元素添加了一个 week 属性和一个 date 属性。第 15 行使用 fxDate 生成 date 对象；第 16 行中的 date 对象的 getDay 方法返回 0～6，代表星期，再根据常数数组 weekArray 得到对应的星期；第 17 行截取 fxDate 的月日部分。

编译运行后，可以查看未来 7 天的天气预报。

至此，天气预报小程序的代码编写完成。

项目小结

本项目介绍了 Picker 组件的多种应用模式，如何在微信小程序中读取外部网站数据并显示在页面中，以及 GET 和 POST 两种读取数据的技术。在此基础上制作了一个功能完备的天气预报微信小程序，读取和风天气网站的天气数据显示在小程序中。学习本项目后，读者可以熟练调用各种 Web API 读取所需的数据。

习题

一、判断题

1．使用 wx.request 只能发起 HTTPS 请求。　（　　）

2．使用 wx.request 发起 GET 请求，不需要指定 method 属性。　（　　）

3．使用省市区选择器时需要指定数据源。　（　　）

二、选择题

1．小程序向服务器发送 POST 请求，需要指定________为 post。

A．data

B．url

C．request

D．method

2．Picker 组件中表示选择的数组当前下标的是________________。

A．range

B．index

C．value

D．item

3．下列关于 picker 说法错误的是（　　）。

A．mode=multiSelector 为多列选择器

B．mode=time 为日期选择器

C．mode=region 为省市区选择器

D．mode=selector 为普通选择器

三、填空题

1．当 picker 标签中的 mode 属性为___________时，Picker 组件为省市区选择器。

2．wx.request 接口调用时需要先设置请求参数和回调函数，调用失败的回调函数是_______。

3．调用 wx.request 接口时，必须指定的一个参数是________。

四、编程题

参照和风天气的开发文档，编程实现查找空气质量，并显示在页面中。显示的内容至少包括 PM10、PM2.5、二氧化氮、二氧化硫、一氧化碳、臭氧。

项目四　用户注册

任务一　使用 ThinkPHP 搭建服务器

4.1.1　小程序传统开发模式简介

小程序用户要提交注册信息到后台数据库，或者从后台数据库获取用户资料并显示在小程序页面，这些过程都需要 Web 服务器、数据接口和后台数据库的参与，它们可以采用微信的云服务，也可以用户自己搭建。前者称为小程序云开发模式，后者称为小程序传统开发模式。由于腾讯的云开发模式需要收费，因此本教材重点介绍传统开发模式的实现。

搭建传统开发模式的技术方案有很多，下面以比较经典的使用 PHPStudy 搭建 PHP Web 服务器和 MySQL 数据库，使用 ThinkPHP 框架搭建数据接口为例来进行介绍。

4.1.2　传统开发模式的环境搭建

首先下载 phpstudy_x64_8.1.1.3.exe 安装程序。PHPStudy 是一款 PHP 环境集成包，内含 PHP 语言运行环境、Apache 和 Nginx 的 Web 服务器、MySQL 数据库及其管理工具 phpMyAdmin，可以一次性安装，无须配置即可使用，是非常方便、好用的 PHP 调试环境。下载完毕后直接双击安装程序进行安装，安装界面如图 4.1.1 所示。之后就可以一键安装 PHP 语言运行环境、Apache 的 Web 服务器、MySQL 数据库等软件和中间件。

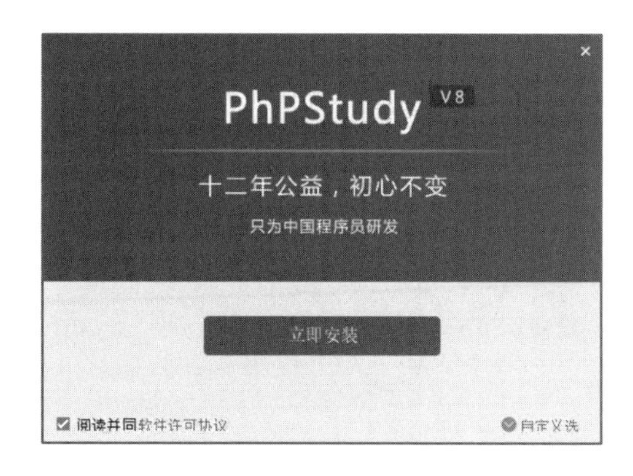

图 4.1.1　phpstudy_x64_8.1.1.3.exe 安装界面

4.1.3　安装 ThinkPHP 6

扫一扫

微课：搭建 ThinkPHP 环境

ThinkPHP 6 在前端小程序和后台数据库之间起到桥梁的作用。

1．安装 Composer

ThinkPHP 是一款开源的 PHP 框架，利用它的控制器和模型，可以更好、更安全地进行数据库的增删改查等各项操作。Windows 系统下的 ThinkPHP 6 以上版本必须通过 Composer 方式安装和更新，因此在 Windows 中需要先下载 Composer-Setup.exe 并安装运行。图 4.1.2 所示是 Composer 的安装界面。

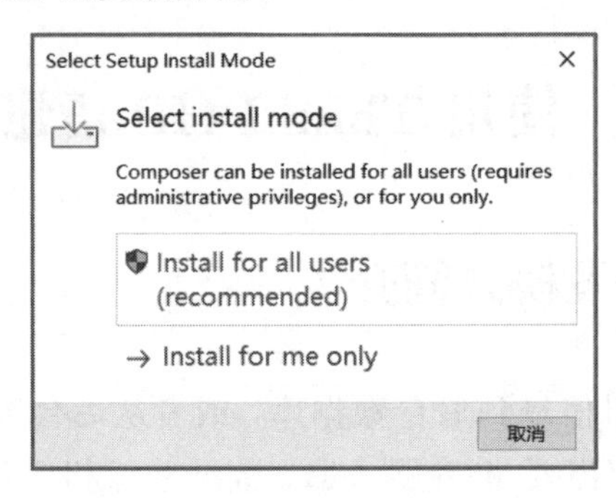

图 4.1.2　Composer 的安装界面

安装时，PHP 的运行文件夹选择安装 PHPStudy 时自动安装 PHP 的文件夹：C:\phpstudy_pro\Extensions\php\php7.3.4nts，如图 4.1.3 所示（图中的 PHPStudy 安装在 D 盘），先单击“Browse”按钮，再选择文件夹。如果安装 PHPStudy 时修改了安装文件夹，则相应地修改 PHP 的运行文件夹。

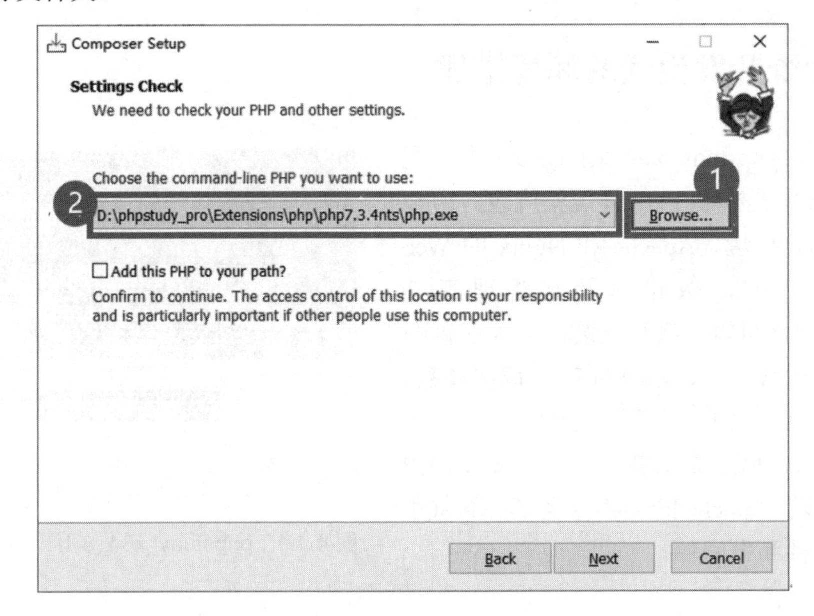

图 4.1.3　选择 PHP 文件夹

选中“Add this PHP to your path?”复选框，单击“Next”按钮继续安装。其他界面直接单击“Next”按钮即可。

2. 安装 ThinkPHP 6，项目名为 miniprog

打开资源管理器，进入 D:\phpstudy_pro\WWW 文件夹，如图 4.1.4 所示。

扫一扫

微课：创建数据库

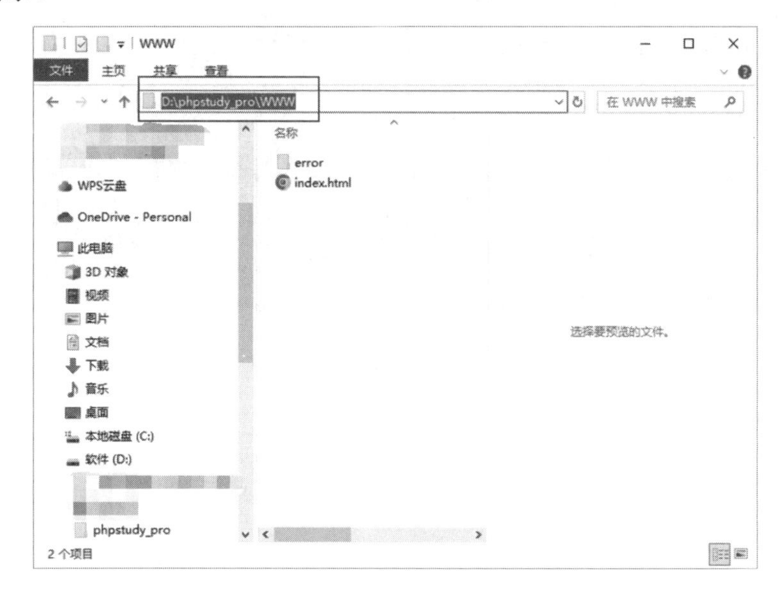

图 4.1.4　进入文件夹

在地址栏输入 cmd，按回车键，如图 4.1.5 所示，打开控制台界面，并进入 D:\phpstudy_pro\WWW 文件夹，该文件夹是 PHPStudy 默认的 Web 文件夹，如图 4.1.6 所示。

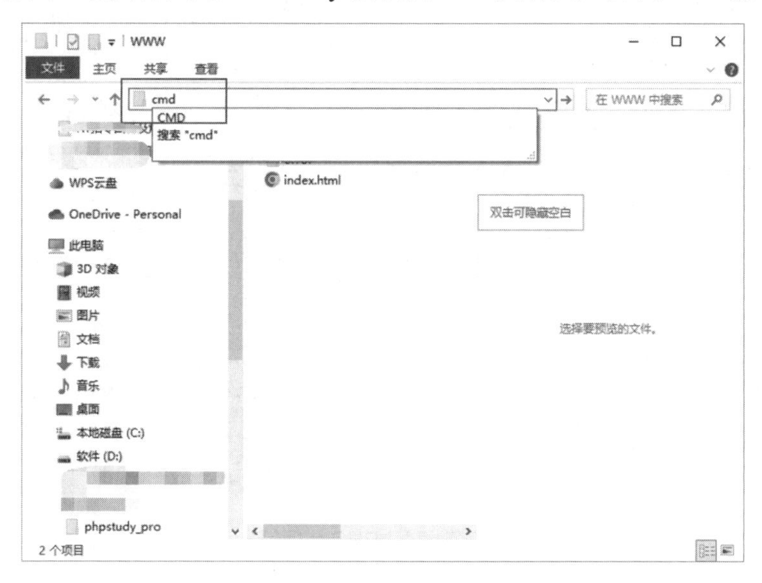

图 4.1.5　输入 cmd

图 4.1.6　控制台界面

依次在控制台输入以下指令：

```
composer config -g repo.packagist composer https://packagist.phpcomposer.com
composer config -g repo.packagist composer https://mirrors.aliyun.com/composer
composer create-project topthink/think miniprog
```

第 1 行和第 2 行用于配置中国镜像地址，任选一行运行；第 3 行用于创建 ThinkPHP 6，项目名称为 miniprog，如图 4.1.7 所示。

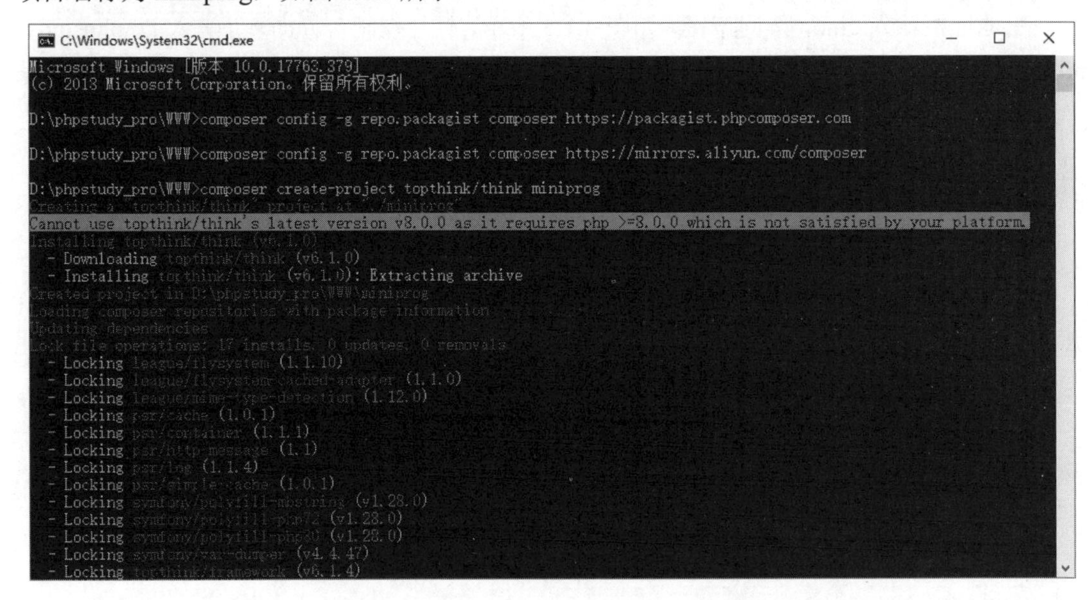

图 4.1.7　安装 ThinkPHP 6 项目 miniprog

命令窗口出现“Succeed!”提示后，说明 ThinkPHP 安装成功，如图 4.1.8 所示。

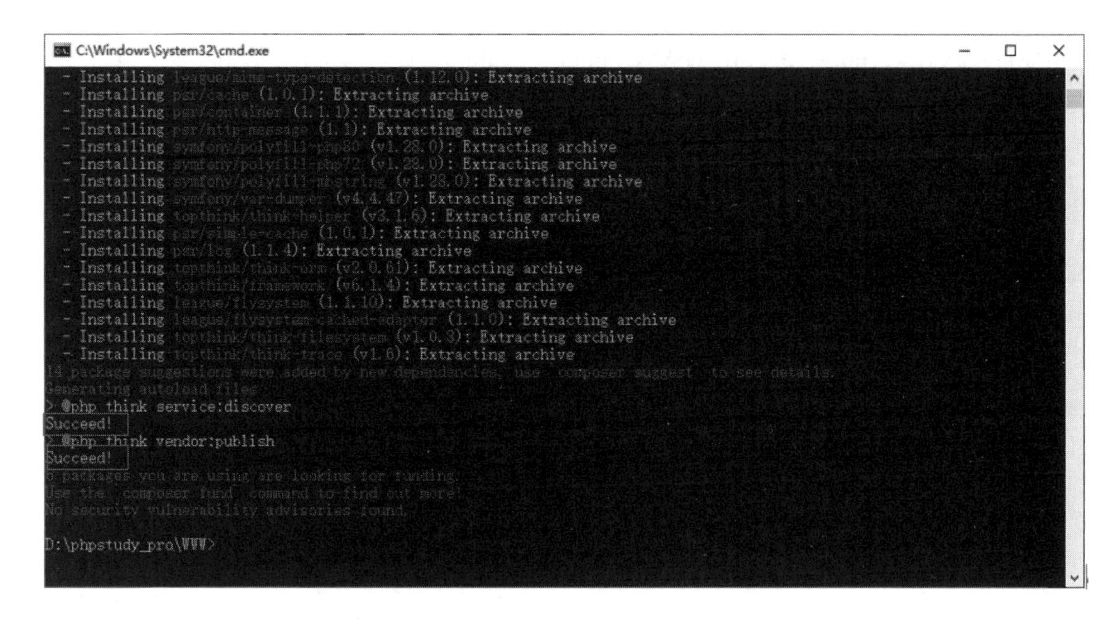

图 4.1.8　安装成功

进入 miniprog 文件夹，运行以下指令：

```
cd miniprog
composer update topthink/framework
```

第 1 行用于进入 miniprog 文件夹，第 2 行用于更新 ThinkPHP 6 核心。

3．配置 localhost 网站

运行 PHPStudy，如图 4.1.9 所示，在界面左侧选择“网站”，单击右侧的“管理”按钮，在下拉菜单中选择“修改”选项。

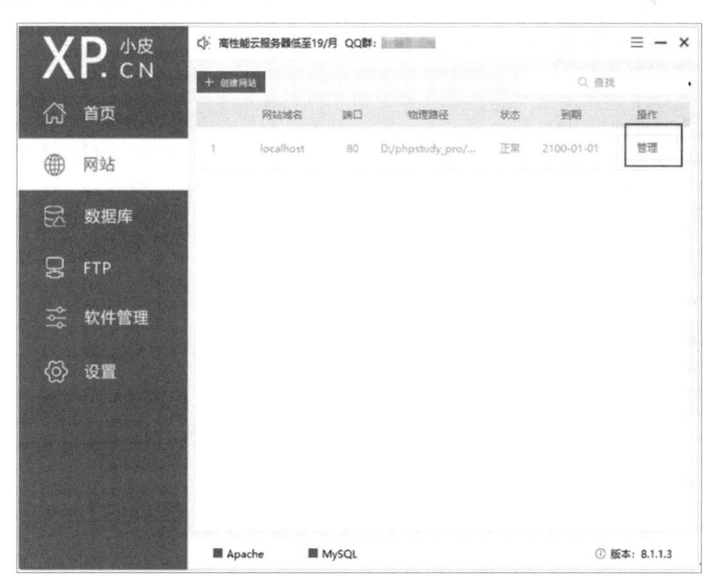

图 4.1.9　配置 localhost 网站

根目录选择 D:/phpstudy_pro/WWW/miniprog/public，如图 4.1.10 所示。

网站
基本配置 高级配置 安全配置 错误页面 伪静态 其他
域名 localhost
第二域名
端口 http https 80
根目录 D:/phpstudy_pro/WWW/miniprog/public 浏览
创建环境 创建FTP 创建数据库 同步hosts 生产环境
程序类型 PHP
PHP版本 php7.3.4nts 到期日期 2100-01-01
备注 localhost站点，请勿随意修改当前站点名称
确认 取消

图 4.1.10　配置根目录

单击“确认”按钮配置完成。如图 4.1.11 所示，在界面左侧选择“首页”，单击 Apache2.4.39 右侧的“启动”按钮启动 Apache 服务器。

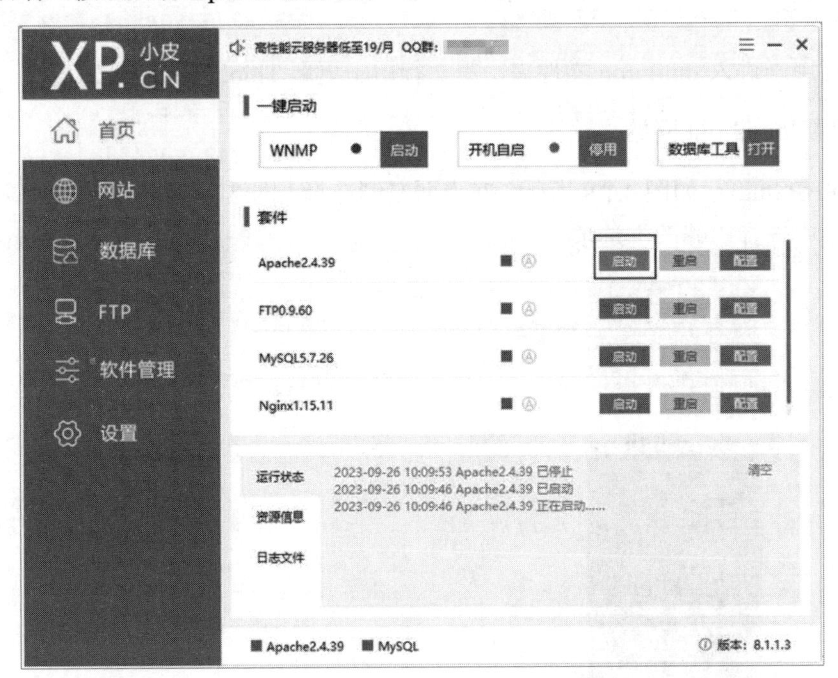

图 4.1.11　启动 Apache 服务器

打开浏览器，在地址栏输入 localhost，弹出 ThinkPHP 6 欢迎页面，如图 4.1.12 所示，说明安装成功。

图 4.1.12　ThinkPHP 6 欢迎页面

4．安装 MySQL 数据库开发工具

运行 PHPStudy，在界面左侧选择“软件管理”，在右侧找到 SQL_Front5.3 和 phpMyAdmin4.8.5，可以任选其中之一安装，也可以都安装，它们都是 MySQL 的图形化开发工具，单击对应的“安装”按钮一步步进行安装，如图 4.1.13 所示。

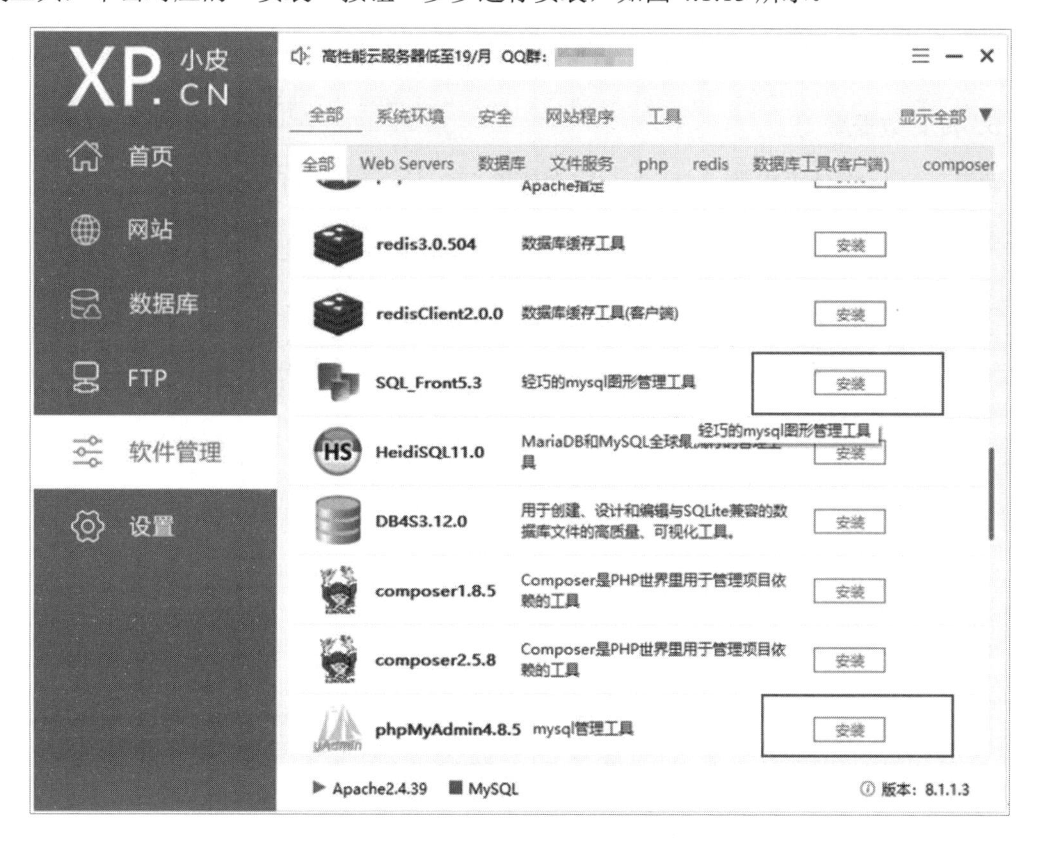

图 4.1.13　安装 MySQL 数据库开发工具

若安装 phpMyAdmin4.8.5，安装到如图 4.1.14 所示页面时，勾选“选择”复选框。

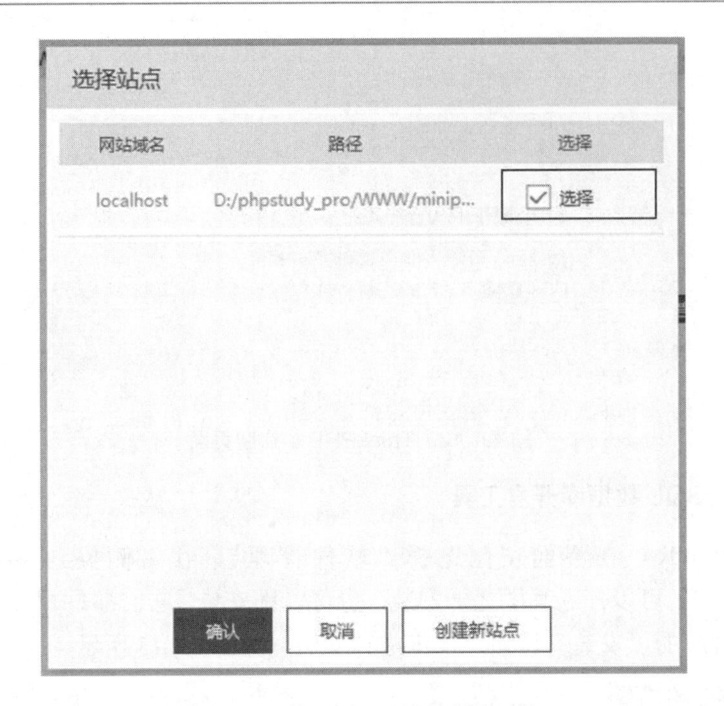

图 4.1.14　勾选“选择”复选框

5．配置数据库

将 miniprog 文件夹下的.example.env 文件重命名为.env，打开文件并进行以下修改：

```
[DATABASE]
TYPE = mysql
HOSTNAME = 127.0.0.1
DATABASE = miniprog
USERNAME = root
PASSWORD = root
HOSTPORT = 3306
CHARSET = utf8
DEBUG = true
```

修改 config/database.php 文件如下：

```
    'mysql' => [
        // 数据库类型
        'type'            => env('database.type', 'mysql'),
        // 服务器地址
        'hostname'        => env('database.hostname', '127.0.0.1'),
        // 数据库名
        'database'        => env('database.database', 'miniprog'),
        // 用户名
        'username'        => env('database.username', 'root'),
        // 密码
        'password'        => env('database.password', 'root'),
```

```
        // 端口
        'hostport'        => env('database.hostport', '3306'),
```

6. 配置局域网可以访问 MySQL

安装 MySQL 后，只能在本机访问数据库服务器，如果在局域网访问 MySQL，就需要进行以下配置。

进入 C:\phpstudy_pro\Extensions\MySQL5.7.26\bin，运行 cmd，执行以下指令：

```
mysql -u root -p
```

输入默认密码 root 后，进入 MySQL 命令行，如图 4.1.15 所示。

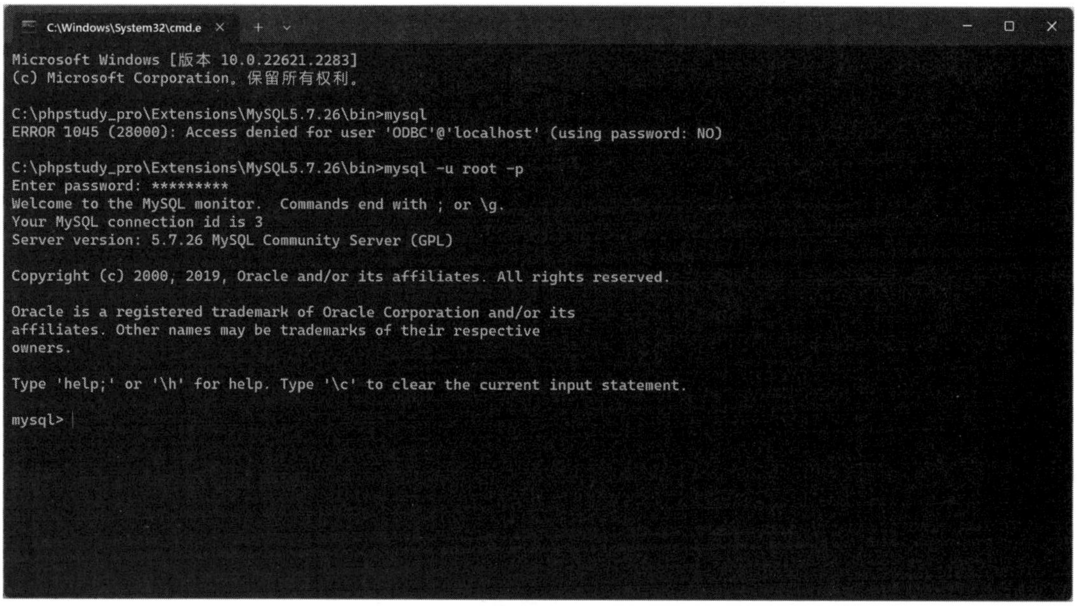

```
C:\Windows\System32\cmd.e
Microsoft Windows [版本 10.0.22621.2283]
(c) Microsoft Corporation。保留所有权利。

C:\phpstudy_pro\Extensions\MySQL5.7.26\bin>mysql
ERROR 1045 (28000): Access denied for user 'ODBC'@'localhost' (using password: NO)

C:\phpstudy_pro\Extensions\MySQL5.7.26\bin>mysql -u root -p
Enter password: *********
Welcome to the MySQL monitor.  Commands end with ; or \g.
Your MySQL connection id is 3
Server version: 5.7.26 MySQL Community Server (GPL)

Copyright (c) 2000, 2019, Oracle and/or its affiliates. All rights reserved.

Oracle is a registered trademark of Oracle Corporation and/or its
affiliates. Other names may be trademarks of their respective
owners.

Type 'help;' or '\h' for help. Type '\c' to clear the current input statement.

mysql>
```

图 4.1.15　进入 MySQL 命令行

依次输入以下指令（注意不能漏掉分号）：

```
use mysql
update user set host='%' where user='root';
flush privileges;
GRANT ALL PRIVILEGES ON *.* TO 'root'@'%' IDENTIFIED BY 'root' WITH GRANT
OPTION;
flush privileges;
```

第 1 条和第 2 条语句用于授权用户 root 可以通过任意主机访问，第 3 条和第 4 条语句用于任意主机以用户 root 和密码 root 连接到 MySQL 服务器。

配置成功后，如果手机与电脑连接同一个 Wi-Fi，则它们处于同一个局域网中，使用手机也可以访问电脑的 MySQL 服务器。

7．优化访问链接，不需要在 url 中加入 index.php

打开 C:/phpstudy_pro/WWW/miniprog/public/.htacces 文件，添加以下代码：

```
<IfModule mod_rewrite.c>
RewriteEngine on
RewriteCond %{REQUEST_FILENAME} !-d
RewriteCond %{REQUEST_FILENAME} !-f
RewriteRule ^(.*)$ index.php [L,E=PATH_INFO:$1]
</IfModule>
```

任务二　了解微信小程序登录流程

微信小程序的登录流程在小程序开发文档中有比较详细的说明。

微信小程序登录流程图如图 4.2.1 所示，主要涉及小程序、开发者服务器和微信接口服务三个对象，分为以下十步。

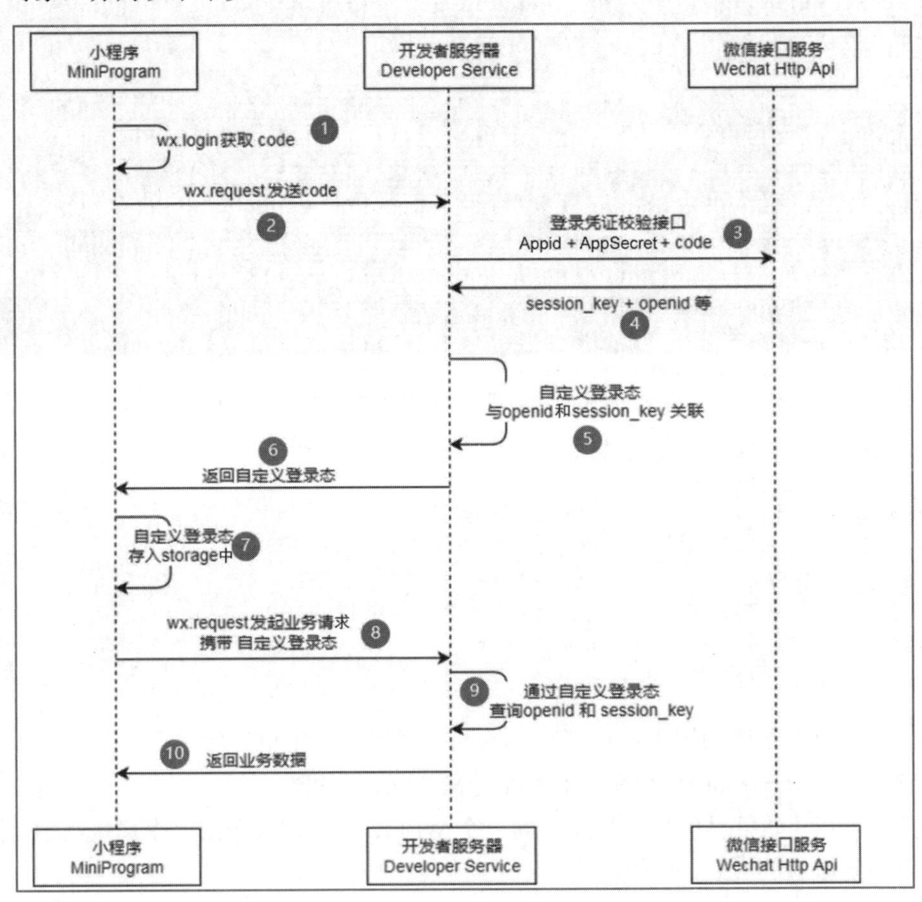

图 4.2.1　微信小程序登录流程图

（1）小程序调用 wx.login 接口获取 code。

（2）小程序把获得的 code 发送给开发者服务器。

（3）开发者服务器把（2）中获得的 code 和小程序对应的 AppID、AppSecret 发送给微信接口服务。

（4）微信接口服务收到 AppID、AppSecret 后返回 session_key 和 openid 给开发者服务器。

（5）开发者服务器利用 session_key 和 openid 生成用户自定义登录态并保存。

（6）开发者服务器把用户自定义登录态返回给小程序。

（7）小程序把开发者服务器返回的用户自定义登录态保存到 storage 中。

（8）用户登录后，当需要向服务器请求服务或获取数据时，小程序把 storage 中保存的用户自定义登录态作为 wx.request 接口中的一个参数发送给开发者服务器。

（9）开发者服务器接收到小程序发送的用户自定义登录态后与保存的用户自定义登录态进行比对，如果登录态是有效的，则返回正常的业务数据给小程序；如果登录态是无效的，则拒绝小程序的业务请求。

（10）开发者服务器发送业务数据或拒绝业务请求的提示信息给小程序。

任务三　使用 ThinkPHP 实现微信登录流程

4.3.1　使用 MySQL 创建表格 token

使用任意工具新建数据库 miniprog，在数据库中创建表格 token，字段如表 4.3.1 所示。

表 4.3.1　token 表格中的字段

字段名	类型	长度	非空	备注
id	int		not null	主键，自动增长
openid	varchar	255	not null	
token	varchar	255	not null	
create_time	int		not null	

对应的创建表格的 sql 代码如下：

```
1	DROP TABLE IF EXISTS 'token';
2	CREATE TABLE 'token'  (
3	  'id' int(11) NOT NULL AUTO_INCREMENT,
4	  'openid' varchar(255) CHARACTER SET utf8 COLLATE utf8_unicode_ci
	NOT NULL,
5	  'token' varchar(255) CHARACTER SET utf8 COLLATE utf8_unicode_ci
	NOT NULL,
6	  'create_time' int(11) NOT NULL,
7	  PRIMARY KEY ('id') USING BTREE
```

```
8	    ) ENGINE = MyISAM AUTO_INCREMENT = 12 CHARACTER SET = utf8 COLLATE
= utf8_unicode_ci ROW_FORMAT = Dynamic;
```

4.3.2 实现微信登录流程

扫一扫

微课：用户注册-登录

打开 app/controller/index.php 文件，在代码“use app\BaseController;”的后面补充以下代码：

```
1	use think\facade\Db;
2	use think\facade\Filesystem;
```

再增加 uuid、send_get 和 login 三种方法：

```
1	    private function uuid()
2	    {
3	        $chars = md5(uniqid(mt_rand(), true));
4	        $uuid = substr($chars, 0, 8) . '-'
5	            . substr($chars, 8, 4) . '-'
6	            . substr($chars, 12, 4) . '-'
7	            . substr($chars, 16, 4) . '-'
8	            . substr($chars, 20, 12);
9	        return $uuid;
10	    }
11	    private function send_get($url, $data)
12	    {
13	        $query = http_build_query($data);
14	        $result = file_get_contents($url . '?' . $query);
15	        return $result;
16	    }
17	    public function login()
18	    {
19	        if (request()->isPost()) {
20	            $param = request()->param();
21	            $code = $param['code'];
22	            $params = array(
23	                'appid' => '',
24	                'secret' => '',
25	                'js_code' => $code,
26	                'grant_type' => 'authorization_code'
27	            );
28	            $res = $this->send_get('https://api.weixin.qq.com/sns/
jscode2session', $params);
29	            //获取微信公众平台的 openid
30	            $data = json_decode($res, true);
31	            if (isset($data['openid'])) {
32	                $token = $this->uuid();
```

```
33                    Db::table('token')->insert([
34                        'openid' => $data['openid'],
35                        'token' => $token,
36                        'create_time' => time()
37                    ]);
38                    return json([
39                        'err_code' => 0,
40                        'token' => $token,
41                        'openid' => $data['openid']
42                    ]);
43                } else {
44                    return json([
45                        'err_code' => 1,
46                        'err_msg' => '获取openid失败'
47                    ]);
48                }
49            } else {
50                return 'login';
51            }
52        }
```

方法 uuid 生成 uuid 字符串作为 token，方法 send_get 用于发送 GET 请求，这两种方法的实现细节超出了本教材的讲解范围。

第 22～27 行是准备调用微信接口服务的参数。在第 23 行需要输入自己的 AppID；在第 24 行需要输入自己的 AppSecret，如果不知道自己的 AppSecret，则进入微信公众平台，登录后选择“开发|开发管理|开发设置”命令，如图 4.3.1 所示。

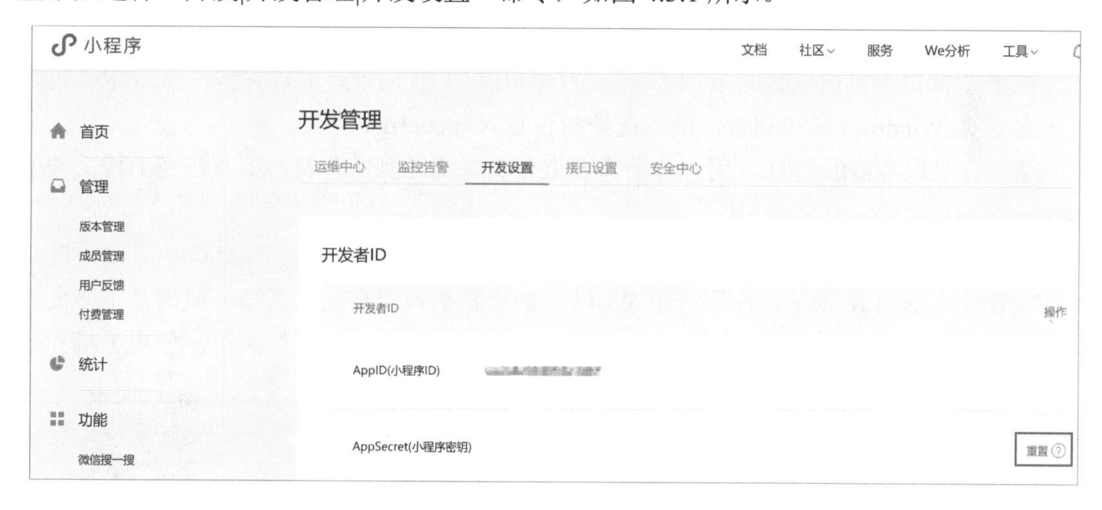

图 4.3.1 查看 AppSecret

如果生成过 AppSecret，则直接使用；如果没有生成过 AppSecret，或者忘记了，则单击“重置”按钮生成新的 AppSecret。如果重置 AppSecret，则原来的 AppSecret 自动失效。

第 28 行用于调用微信接口服务。调用成功后获得的 openid 保存在$data['openid']中。第 32 行生成新的 uuid 作为 token。第 33～37 行在数据库的 token 表格中插入一条记录。第 38～42 行返回 json 格式的 openid 和 token 作为应答信息。第 44～47 行返回错误信息。

任务四　实现微信登录

打开微信开发者工具，新建项目，名称为 register，使用自己注册的 AppID，不使用模板。打开 app.json 文件，修改“navigationBarTitleText”为“用户注册”。

复制资源文件夹中 register 下的 images 文件夹到新建的 register 微信小程序文件夹，使该文件夹与 pages 文件夹并列，其中包含一个 avatar.png 图片作为默认的头像文件。

4.4.1　添加服务器 IP 配置文件

在小程序根文件夹新建文件夹 utils，可以在微信开发者工具“资源管理器”底部的空白处右击，在弹出的快捷菜单中选择“新建文件夹”命令。

在 utils 文件夹中新建文件 settings.js，并输入以下代码：

```
1    const ip = '***.***.***.***'
2    module.exports = {
3      server_url: 'http://' + ip + '/index/',
4      file_url: 'http://' + ip + '/storage/'
5    }
```

在第 1 行输入自己电脑的 IP 地址。要查看电脑的 IP 地址，可以先输入 win+R，再输入 cmd 进入 Windows 的控制台，最后在控制台输入 ipconfig。

第 3 行是服务器的 URL。用户注册需要上传头像文件到服务器，第 4 行是上传文件的 URL。

如果只是在微信开发者工具里访问服务器，可以将 IP 地址设置为 127.0.0.1。如果将 IP 地址设置成电脑的 IP 地址，在同一个 Wi-Fi 下的手机也可以访问服务器，前提是电脑允许局域网访问本机端口。设置好之后，单击微信开发者工具的“预览”按钮，使用手机扫码就可以运行小程序，并连接到服务器。

扫一扫

微课：用户注册-登录代码 1

扫一扫

微课：用户注册-登录代码 2

4.4.2　自动执行微信登录流程

小程序运行后，需要自动执行微信登录流程，因此需要将代码添加到 app.js 文件中。

打开 app.js 文件，将其修改为以下代码：

```
1    const settings = require('/utils/settings.js')
```

```
2
3    App({
4      onLaunch: function () {
5        let token = this.globalData.token
6        if (!token) {
7          token = wx.getStorageSync('token')
8          if (token) {
9            this.globalData.token = token
10         } else {
11           this.login()
12         }
13       }
14     },
15
16     login: function () {
17       wx.login({
18         success: res => {
19           console.log('login code: ' + res.code)
20           console.log(settings.server_url + 'login?code='+res.code)
21           wx.request({
22             url: settings.server_url + 'login',
23             method: 'post',
24             data: { code: res.code },
25             success: res => {
26               console.log(res.data)
27               // 将token保存为公共数据（用于在多页面中访问）
28               this.globalData.token = res.data.token
29               // 将token保存到本地缓存中（再次打开小程序时无须重新获取token）
30               wx.setStorage({ key: 'token', data: res.data.token })
31             },
32             fail: err => {
33               console.error(err)
34             }
35           })
36         }
37       })
38     },
39     globalData: {
40       token: null
41     }
42   })
```

第 1 行引入添加的配置文件。小程序启动后，先执行 App 的 onLaunch 方法。

登录成功后会将服务器返回的 token 保存在本地缓存中，同时保存在 App 的 globalData 中。第 6 行检查 globalData 中是否有 token，第 7 行读取本地缓存中的 token，第 8 行和第 9 行检查在本地缓存中是否读到了 token。第 11 行是没有读取到 token，代表没有登录过，调用 login 方法进行登录。

第 17 行调用 wx.login 方法获取用户登录凭证（有效期为五分钟），即 code，调用成功后通过 res.code 返回用户登录凭证。第 21～26 行请求通过本项目任务三中的服务器实现登录，请求成功后通过 res.data.token 返回服务器生成的 token。第 28 行将 token 保存到 globalData 中，第 29 行将 token 保存到本地缓存中。

wx.setStorage 将数据存储在本地缓存指定的 key 中。除非用户主动删除或因存储空间不足被系统清理，否则本地缓存中的数据一直可用。单个 key 允许存储的最大数据长度为 1MB，所有数据的存储上限为 10MB。

wx.getStorageSync 从本地缓存中同步获取指定 key 的内容。

任务五　使用小程序常用表单组件

表单的作用是获取用户的输入并提交给后台服务器进行处理。小程序的表单组件主要包括单行文本输入框、多行文本输入框、单选按钮、复选按钮、开关按钮、滑动选择器、Picker 选择器等。下面介绍这些表单组件的属性和用法（项目三中已经介绍了 Picker 组件的使用，本项目不再重复介绍）。

4.5.1　单行文本输入框

单行文本输入框组件 input 的属性如表 4.5.1 所示。

表 4.5.1　input 组件的属性

属性	类型	默认值	说明
value	string		输入框中的初始内容
type	string	text	input 的类型，有效值为 text（文本输入键盘）、number（数字输入键盘）、idcard（身份证输入键盘）、digit（带小数点的数字输入键盘）
password	boolean	False	是否为密码类型
placeholder	string		输入框为空时的占位符
disabled	boolean	False	是否禁用
bindinput	eventhandle		当键盘输入时触发 input 事件并绑定处理函数，event.detail.value 获取输入框中的内容
bindfocus	eventhandle		输入框聚焦时触发 focus 事件并绑定处理函数，event.detail.value 获取输入框中的内容

续表

属性	类型	默认值	说明
bindblur	eventhandle		输入框失去焦点时触发 blur 事件并绑定处理函数，event.detail.value 获取输入框中的内容
bindconfirm	eventhandle		单击“完成”按钮时触发 confirm 事件并绑定处理函数，event.detail.value 获取输入框中的内容

input 组件的使用示例如图 4.5.1 所示。

图 4.5.1 input 组件的使用示例

要实现这个效果，可以打开微信开发者工具，使用测试账号新建一个微信小程序，不使用模板。打开 index.wxml 文件，删除原来的代码，输入以下代码：

```
1    <view class="container">
2        <view style="font-weight:bold;margin-top: 20px;">普通文本框</view>
3        <input placeholder="请输入文本" style="border-bottom: green;border-bottom-width:1px ;border-bottom-style: solid;" bindblur="getBlurText" />
4      <view style="font-weight:bold;margin-top: 20px;">设置最大输入长度的文本框</view>
5        <input  placeholder="请输入文本" maxlength="8" style="border-bottom: green;border-bottom-width:1px ;border-bottom-style: solid;" bindinput="getInputingText" />
6      <view style="font-weight:bold;margin-top: 20px;">输入数字的文本框</view>
7        <input type="digit" placeholder="请输入数字" style="border-bottom:green;border-bottom-width:1px ;border-bottom-style: solid;" bindblur="getBlurText"/>
8      <view style="font-weight:bold;margin-top: 20px;">输入密码的文本框</view>
9        <input type="text" password="true" placeholder="请输入密码" style="border-bottom: green;border-bottom-width:1px ;border-bottom-style: solid;" bindblur="getBlurText"/>
10     <view style="font-weight:bold;margin-top: 20px;">输入身份证的文本框</view>
```

```
11      <input type="idcard" placeholder="请输入身份证号码" style="border-bottom: green;border-bottom-width:1px ;border-bottom-style: solid;" bindinput="getInputingText"/>
12    </view>
```

第3行插入了一个input组件的代码，bindblur="getBlurText"表示使用js中的getBlurText函数处理失去焦点事件；第5行中的input组件通过maxlength="8"设置最大输入长度为8个字符；第7行中的type="digit"设置只能输入数字类型的值（在微信开发者工具的模拟器中无效，在手机上运行才有效）；第9行中的type="text"设置以实心圆点代替原来的输入字符；第11行中的type="idcard"设置只能输入身份证号码，若输入不符合要求将提示重新输入。

打开index.js文件，删除原来的代码，输入以下代码：

```
1    Page({
2      getBlurText(e){
3        console.log("输入的文本是："+e.detail.value)
4      },
5      getInputingText(e){
6        console.log("正在输入的文本是："+e.detail.value)
7      }
8    })
```

这段代码定义了一个getBlurText函数处理失去焦点事件，getInputingText函数则是正在输入事件对应的处理函数。

4.5.2 多行文本输入框

当表单中要输入的文本内容比较多时，可以使用textarea多行文本框组件。多行文本框组件的主要属性如表4.5.2所示。

表4.5.2　多行文本框组件的主要属性

属性	类型	默认值	说明
value	string		输入框中的初始内容
placeholder	string		输入框为空时的占位符
disabled	boolean	False	是否禁用
maxlength	number	140	最大输入长度，设置为-1时不限制最大输入长度
auto-height	boolean	False	是否自动增高，设置auto-height属性时，style.height不生效
bindinput	eventhandle		当键盘输入时触发input事件并绑定处理函数，event.detail.value获取输入框中的内容
bindfocus	eventhandle		输入框聚焦时触发focus事件并绑定处理函数，event.detail.value获取输入框中的内容

续表

属性	类型	默认值	说明
bindblur	eventhandle		输入框失去焦点时触发 blur 事件并绑定处理函数，event.detail.value 获取输入框中的内容
bindconfirm	eventhandle		单击“完成”按钮时触发 confirm 事件并绑定处理函数，event.detail.value 获取输入框中的内容

多行文本框组件的使用示例如图 4.5.2 所示。

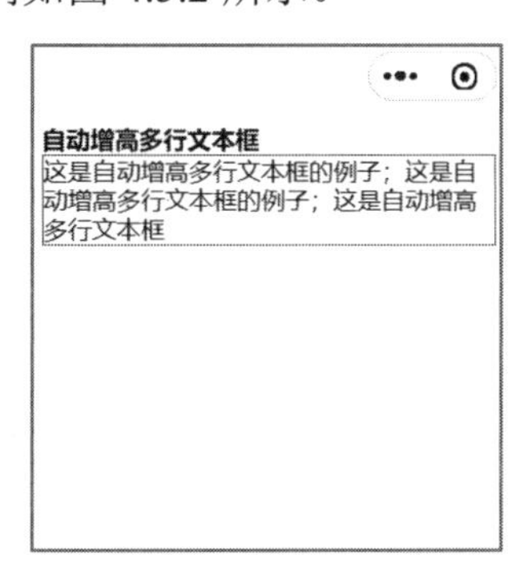

图 4.5.2　多行文本框组件的使用示例

要实现这个效果，可以打开微信开发者工具，使用测试账号新建一个微信小程序，不使用模板。打开 index.wxml 文件，删除原来的代码，输入以下代码：

```
1     <view class="container">
2       <view style="font-weight:bold;margin-top: 20px;">自动增高多行文本框
  </view>
3       <textarea placeholder="请输入多行文本" style="border-color: green;
  border-width:1px ;border-style: solid;" bindblur="getBlurText" auto-
  height="true"/>
4
5     </view>
```

第 3 行中的 textarea 标签定义了一个多行文本输入框，auto-height 属性为“true”，使输入框具有根据文本自动调整高度的特点。

打开 index.js 文件，删除原来的代码，输入以下代码：

```
1    Page({
2      getBlurText(e){
3        console.log("输入的文本是："+e.detail.value)
4      },
5    })
```

4.5.3　单选按钮

radio 组件是单选按钮组件，通常和 radio-group 组件配合使用，主要应用在二选一的场景中，如性别选择、是否选择等。radio 组件的属性如表 4.5.3 所示。

表 4.5.3　radio 组件的属性

属性	类型	默认值	说明
value	string		<radio/>标识。当该<radio/>选中时，<radio-group/>的 change 事件会携带<radio/>的 value
disabled	boolean	False	是否禁用
checked	boolean	False	当前是否选中。可用来设置默认选中
color	color		radio 的颜色，同 CSS 中的 color

radio 组件和 radio-group 组件的使用示例如图 4.5.3 所示。

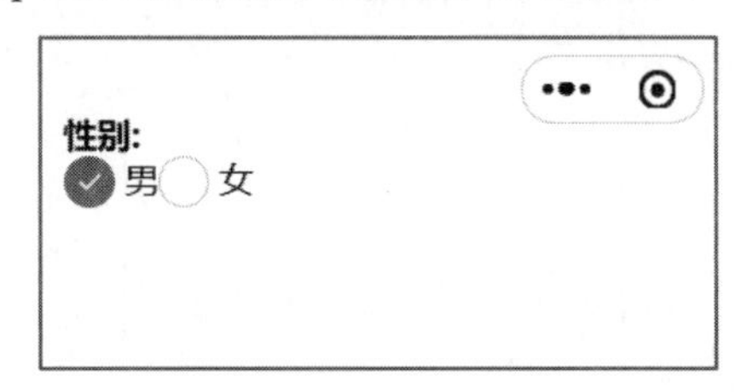

图 4.5.3　radio 组件和 radio-group 组件的使用示例

要实现这个效果，可以打开微信开发者工具，使用测试账号新建一个微信小程序，不使用模板。打开 index.wxml 文件，删除原来的代码，输入以下代码：

```
1    <view class="container">
2      <view style="font-weight:bold;">性别:</view>
3        <radio-group name="radio-group" bindchange="radiochange">
4          <label><radio value="男" checked="true"/>男</label>
5          <label><radio value="女"/>女</label>
6        </radio-group>
7      </view>
8    </view>
```

第 3 行定义了一个 radio-group 组件，它把两个 radio 组件编为一组，操作时只能选择其中之一；第 4 行中的 checked="true"表示所在的 radio 组件默认处于被选中状态。

打开 index.js 文件，删除原来的代码，输入以下代码：

```
1    Page({
2      radiochange(e){
3        console.log("选择的性别是: "+e.detail.value)
4      }
5    })
```

第 2 行和第 3 行代码定义了一个 radio-group 组件中被选中的 radio 组件发生变化事件对应的处理函数 radiochange。

4.5.4　checkbox 复选按钮

checkbox 组件是复选按钮组件，通常和 checkbox-group 组件配合使用，主要应用在具有两个以上选项的场景，如兴趣爱好、喜欢的明星等。checkbox 组件的属性如表 4.5.4 所示。

表 4.5.4　checkbox 组件的属性

属性	类型	默认值	说明
value	string		<checkbox/>标识，选中时触发<checkbox-group/>的 change 事件，并携带<checkbox/>的 value
disabled	boolean	False	是否禁用
checked	boolean	False	当前是否选中。可用来设置默认选中
color	color		checkbox 的颜色，同 CSS 中的 color

checkbox 组件和 checkbox-group 组件的使用示例如图 4.5.4 所示。

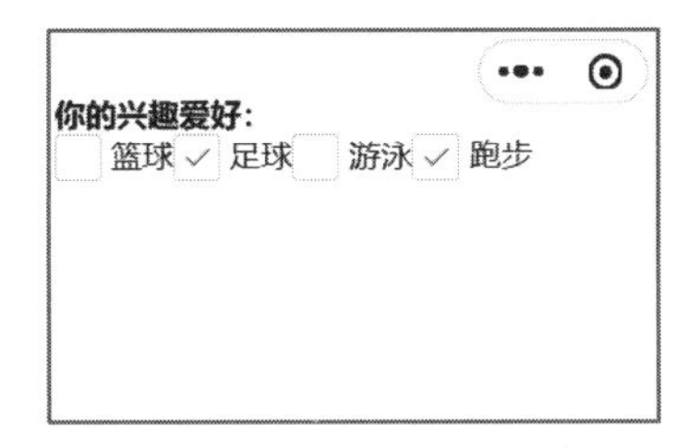

图 4.5.4　checkbox 组件和 checkbox-group 组件的使用示例

要实现这个效果，可以打开微信开发者工具，使用测试账号新建一个微信小程序，不使用模板。打开 index.wxml 文件，删除原来的代码，输入以下代码：

```
1   <view class="container">
2     <view style="font-weight:bold;">你的兴趣爱好：</view>
3       <checkbox-group name="checkbox" bindchange="checkboxChange">
4         <label><checkbox value="篮球"/>篮球</label>
5         <label><checkbox value="足球"/>足球</label>
6         <label><checkbox value="游泳"/>游泳</label>
7         <label><checkbox value="跑步"/>跑步</label>
8       </checkbox-group>
9     </view>
10  </view>
```

checkbox-group 组件把四个 checkbox 组件编为一组，操作时可以选择 0 至多个组件，每个 checkbox 组件都有自己的 value 值。

打开 index.js 文件，删除原来的代码，输入以下代码：

```
1   Page({
2
3     checkboxChange(e){
```

```
4          console.log("你的兴趣爱好是: "+e.detail.value)
5        }
6      })
```

第 3 行中的 checkboxChange 函数是复选框的选择项发生变化事件对应的处理函数。

4.5.5 slider 组件和 switch 组件

slider 组件是滑动选择组件，用于数字类型的值的输入，如输入年龄、体重等。switch 组件是开关组件，应用在是否、有无等二选一的场景中。slider 组件的主要属性如表 4.5.5 所示，switch 组件的主要属性如表 4.5.6 所示。

表 4.5.5　slider 组件的主要属性

属性	类型	默认值	说明
min	number	0	最小值
max	number	100	最大值
step	number	1	步长，取值必须大于 0，并且可以被(max-min)整除
disabled	boolean	False	是否禁用
value	number	0	当前取值
color	color	#e9e9e9	背景条的颜色（使用 backgroundColor）
selected-color	color	#1aad19	已选择的颜色（使用 activeColor）
activeColor	color	#1aad19	已选择的颜色
backgroundColor	color	#e9e9e9	背景条的颜色
show-value	boolean	False	是否显示当前 value
bindchange	eventhandle		完成一次拖动后触发的事件，event.detail = {value: value}

表 4.5.6　switch 组件的主要属性

属性	类型	默认值	说明
checked	boolean	False	是否选中
type	string	switch	样式，有效值为 switch、checkbox
disabled	boolean	False	是否禁用开关
color	string	#04BE02	switch 的颜色，同 CSS 中的 color
bindchange	eventhandle		checked 改变时触发 change 事件，event.detail={ value:checked}

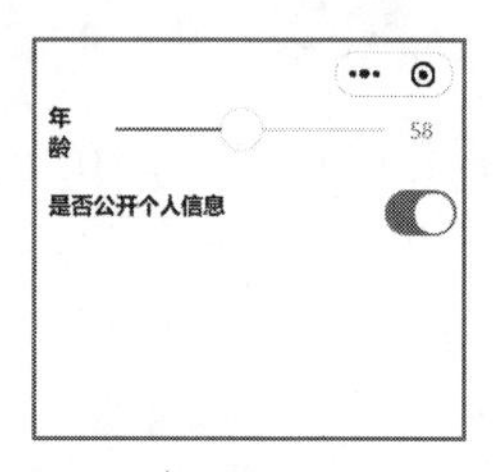

图 4.5.5　slider 组件和 switch 组件的使用示例

下面在同一个示例中说明 slider 组件和 switch 组件的使用方法，如图 4.5.5 所示。

要实现这个效果，可以打开微信开发者工具，使用测试账号新建一个微信小程序，不使用模板。打开 index.wxml 文件，删除原来的代码，输入以下代码：

```
1       <view class="container">
2         <view style="margin:10px;display:flex;justify-content: space-
        between;">
3           <view style="font-weight:bold;">年龄</view>
4             <slider name="slider" min="3" max="120" value="33" style=
        "width: 250px;" show-value bindchange="sliderChange"></slider>
5           </view>
6           <view style="margin:10px;display:flex;justify-content: space-
        between;">
7             <view style="font-weight:bold;">是否公开个人信息</view>
8             <switch name="switch" checked="true" style="right: 10px;"
        bindchange="switchChange"/>
9           </view>
10        </view>
11      </view>
```

第 4 行定义了一个 slider 组件，设置了最小值、最大值和默认值；第 8 行定义了一个 switch 组件，通过 checked="true"设置了默认选中状态。

打开 index.js 文件，删除原来的代码，输入以下代码：

```
1     Page({
2       sliderChange(e){
3         console.log("输入的年龄是："+e.detail.value)
4       },
5       switchChange(e){
6         console.log("是否公开个人信息："+e.detail.value)
7       }
8     })
```

第 2 行定义了一个 slider 组件值发生变化事件对应的处理函数 sliderChange，第 5 行定义了一个 switch 组件状态值发生变化事件对应的处理函数 switchChange。

任务六　制作用户注册小程序

扫一扫

微课：用户注册-注册代码

4.6.1　设计注册页面

打开 index.wxml 文件，删除原来的代码，输入以下代码：

```
1       <view class="container">
2         <view class="avatar">
3           <image src="{{avatar}}" mode="aspectFill" bindtap="changeAvatar">
        </image>
4         </view>
5         <form bindsubmit="submit">
```

```
6          <view>
7            <text>姓名：</text>
8            <input name="name" value="{{name}}" placeholder="请输入姓名" />
9          </view>
10         <view>
11           <text>性别：</text>
12           <radio-group name="gender">
13             <label wx:for="{{['男','女']}}" wx:key="*this">
14               <radio value="{{item}}" checked="{{gender==item?true:
false}}" />
15               {{item}}
16             </label>
17           </radio-group>
18         </view>
19         <view>
20           <text>爱好：</text>
21           <checkbox-group name="hobby">
22             <label wx:for="{{['足球','羽毛球','篮球','乒乓球','排球']}}"
wx:key="*this">
23               <checkbox value="{{item}}" checked="{{array_function.
in_array(hobby, item)?true:false}}" />
24               {{item}}
25             </label>
26           </checkbox-group>
27         </view>
28         <view>
29           <picker name="nationality" bindchange="pickerChange" value=
"{{index}}" range="{{country}}">
30             <view class="picker">
31               国籍：{{country[index]}}
32             </view>
33           </picker>
34         </view>
35         <view>
36           <switch name='is_admin' checked="{{is_admin}}">管理员</switch>
37         </view>
38         <view>
39           <text>您的简介：</text>
40           <textarea name="intro" value="{{intro}}" />
41         </view>
42         <button type="primary" form-type="submit">提交</button>
43       </form>
44     </view>
45     <wxs module="array_function">
46       //从js传入array数组，需要判断的id
47       function in_array(array, item) {
```

```
48        return array.indexOf(item) >= 0
49      }
50      module.exports = {
51        in_array: in_array,
52      }
53    </wxs>
```

本段代码使用了 image、input、radio、checkbox、picker、switch、textarea 等常用表单组件。

第 13～16 行使用了列表渲染，第 14 行中的{{gender==item?true:false}}表示如果 gender 变量等于当前项（男或女），则选中，否则不选中。

第 22～25 行也使用了列表渲染，第 23 行中的{{array_function.in_array(hobby, item)?true:false}}调用了第 45～53 行的微信脚本（WeiXin Script，WXS）。hobby 是供用户选择的爱好，类型是数组。in_array 方法是微信脚本中编写的方法，传入数组和字符串，如果字符串在数组中，则返回 true，否则返回 false。array_function.in_array(hobby, item)?true:false 表示如果当前项 item 包含在 hobby 数组中，则选中该选项，否则不选中。比如，如果字符串“羽毛球”包含在 hobby 数组中，则“羽毛球”选项被选中。

> WXS 是嵌入 WXML 中的脚本。在 WXML 内部嵌入微信脚本，可以提高 WXML 的数据处理能力。从语法上看，WXS 非常接近 JavaScript，但有少量限制。

因为 WXML 内部不直接支持第 48 行的 indexOf 函数，所以要判断当前项在数组中是否无法直接在 WXML 中实现，也不方便在 js 代码中实现，在这种情况下可以使用微信脚本。微信脚本的详细介绍可以参考微信官方文档。

打开 index.wxss 文件，输入以下代码：

```
1   page {
2     font-size: medium;
3   }
4   view {
5     margin-bottom: 20rpx;
6   }
7
8   input {
9     width: 600rpx;
10    margin-top: 10rpx;
11    border-bottom: 2rpx solid #666;
12  }
13
14  label {
15    display: block;
16    margin: 10rpx;
```

```
17      }
18
19      textarea {
20        width: 600rpx;
21        height: 200rpx;
22        margin-top: 10rpx;
23        border: 2rpx solid #666;
24      }
25
26      .avatar {
27        width: 100%;
28        height: 300rpx;
29        background-color:rgb(144, 219, 238);
30        display: flex;
31        justify-content: center;
32        align-items: center;
33      }
34
35      .avatar>image {
36        width: 200rpx;
37        height: 200rpx;
38        border-radius: 50%;
39        border: 10rpx solid rgba(0, 0, 0, 0.1);
40      }
```

4.6.2 获取用户信息

扫一扫

微课：用户注册-用户信息代码

如前文所述，openid 是用户在当前小程序的唯一标识。因此可以将与该用户关联的数据与 openid 绑定，用户登录成功后先获取 openid，再通过它就可以查询到所有与此用户关联的数据。

1. 创建 user 表

用户的注册信息保存在单独的数据库表格中。在数据库 miniprog 中创建表格 user，字段如表 4.6.1。

表 4.6.1　user 表格中的字段

字段名	类型	长度	非空	备注
openid	varchar	100	not null	主键
name	varchar	10		
gender	varchar	2		
hobby	varchar	100		
nationality	varchar	50		

续表

字段名	类型	长度	非空	备注
is_admin	bit	1		
intro	varchar	255		
avatar	varchar	255		

对应的创建表格的 sql 代码如下：

```
1      CREATE TABLE 'user'  (
2        'openid' varchar(100) CHARACTER SET utf8 COLLATE utf8_unicode_ci
   NOT NULL,
3        'name' varchar(10) CHARACTER SET utf8 COLLATE utf8_unicode_ci
   NULL DEFAULT NULL,
4        'gender' varchar(2) CHARACTER SET utf8 COLLATE utf8_unicode_ci
   NULL DEFAULT NULL,
5        'hobby' varchar(100) CHARACTER SET utf8 COLLATE utf8_unicode_ci
   NULL DEFAULT NULL,
6        'nationality' varchar(50) CHARACTER SET utf8 COLLATE utf8_
   unicode_ci NULL DEFAULT NULL,
7        'is_admin' bit(1) NULL DEFAULT NULL,
8        'intro' varchar(255) CHARACTER SET utf8 COLLATE utf8_unicode_ci
   NULL DEFAULT NULL,
9        'avatar' varchar(255) CHARACTER SET utf8 COLLATE utf8_unicode_ci
   NULL DEFAULT NULL,
10       PRIMARY KEY ('openid') USING BTREE
11     ) ENGINE = MyISAM CHARACTER SET = utf8 COLLATE = utf8_unicode_ci
   ROW_FORMAT = Dynamic;
```

2. ThinkPHP 处理用户信息

用户注册时将信息提交到后台服务器保存，同时携带从服务器获取的 token 来表明自己的身份，由服务器进行身份认证。

打开 app/controller/index.php 文件，增加 check_token 和 user_info 两种方法。

```
1       protected $expire = 3 * 60 * 60; //token 的有效期，单位：秒
2        private function check_token($token)
3        {
4            $data = DB::table('token')->where('token', $token)->find();
5            if ($data == null) {
6                return [
7                    'err_code' => 2,
8                    'err_msg' => 'token' . $token . '不存在'
9                ];
10           } else if (time() - $data['create_time'] > $this->expire) {
11               return [
12                   'err_code' => 3,
13                   'err_msg' => 'token' . $token . '已过期'
```

```
14                ];
15            } else {
16                return [
17                    'err_code' => 0,
18                    'openid' => $data['openid']
19                ];
20            }
21        }
22
23        public function user_info()
24        {
25            if (request()->isPost()) {
26                $param = request()->param();
27                $token = $param['token'];
28                $result = $this->check_token($token);
29                if ($result['err_code'] > 0) {
30                    return json($result);
31                }
32                $openid = $result['openid'];
33                $user = DB::table('user')->where('openid', $openid)
->find();
34                if ($user == null) {
35                    return json([
36                        'err_code' => 4,
37                        'err_msg' => 'openid' . $openid . '还没有注册用户'
38                    ]);
39                } else {
40                    return json([
41                        'err_code' => 0,
42                        'user_info' => $user
43                    ]);
44                }
45            } else {
46                return 'user_info';
47            }
48        }
```

第 1 行定义了 token 的有效期是 3 个小时。check_token 方法用于检查 token 是否存在及是否过期，如果存在且不过期，则返回对应的 openid。第 4 行查询数据库表 token，如果查询不到，则代表 token 不存在，第 10 行检查 token 是否过期。

user_info 方法返回用户注册信息。第 33 行使用 openid 查找数据库表 user，如果查询不到，则代表用户没有注册，否则返回用户注册信息。

3. 小程序实现获取用户信息

回到微信开发者工具，打开 index.js 文件，删除原来的代码，输入以下代码：

```
1     const settings = require('../../utils/settings.js')
2
3     const app = getApp()
4     Page({
5       data: {
6         name: '',//input
7         gender: '男',//radio
8         hobby: ['足球', '篮球'],//checkbox
9         nationality: '',//国籍, picker
10        is_admin: true,//是否是管理员, switch
11        intro: '',//简介, textarea
12        avatar: '/images/avatar.png',//头像, image
13        //使用picker
14        country: ['中国', '美国', '加拿大', '意大利', '俄罗斯', '其他'],
15        index: 0,
16      },
17      avatar: '',
18      pickerChange: function (e) {
19        this.setData({
20          index: e.detail.value
21        })
22      },
23      showMessage: function (str) {
24        wx.showToast({
25          title: str,
26          duration: 1500,
27          icon: 'none'
28        })
29      },
30      getUserInfo: function () {
31        let token = app.globalData.token
32        if (!token) {
33          this.showMessage('还没有登录，请稍后重试')
34          app.login()
35          return
36        }
37        wx.request({
38          url: settings.server_url + 'user_info',
39          method: 'post',
40          data: { token: token },
41          success: res => {
42            console.log(res.data)
43            if (res.data.err_code == 0) {
44              const user_info = res.data.user_info
45              this.setData({
46                name: user_info.name,//input
```

```
47                  gender: user_info.gender,//radio
48                  hobby: user_info.hobby.split(','),//checkbox
                    index: this.data.country.indexOf(user_info.nationality),
49  //国籍，picker
50                  is_admin: user_info.is_admin,//是否是管理员，switch
51                  intro: user_info.intro,//简介，textarea
52                  avatar: user_info.avatar && user_info.avatar != '' ?
                      settings.file_url + user_info.avatar : '/images/avatar.
53  png'
54                })
55              } else if (res.data.err_code == 2 || res.data.err_code == 3) {
56                //token 不存在或已过期
57                app.login()
58                this.showMessage('登录信息不存在或已过期，请下拉刷新重试')
59              }
60            },
61            fail: err => {
62              console.error(err)
63            }
64          })
65        },
66        onLoad: function (options) {
67          this.getUserInfo()
68        },
69        onPullDownRefresh: function () {
70          this.getUserInfo()
71        },
72      })
```

为了能够随时刷新用户注册信息，采用常用的下拉刷新重试。在小程序中实现下拉刷新，需要进行配置。打开 index.json 文件，增加配置"enablePullDownRefresh": true，整个文件的代码如下：

```
1   {
2     "usingComponents": {},
3     "enablePullDownRefresh": true
4   }
```

同时，index.js 文件第 69 行中的 onPullDownRefresh 事件用于监听下拉刷新。

页面的 onPullDownRefresh 事件用于监听下拉刷新。要使页面支持下拉刷新事件，需要进行配置，可以在 app.json 文件中进行全局配置，使所有页面都支持下拉刷新事件；也可以在页面的 json 文件中进行配置，如 index.json 文件的配置。app.json 文件的配置方法与页面相同，只需配置在 app.json 文件的 window 选项中即可。

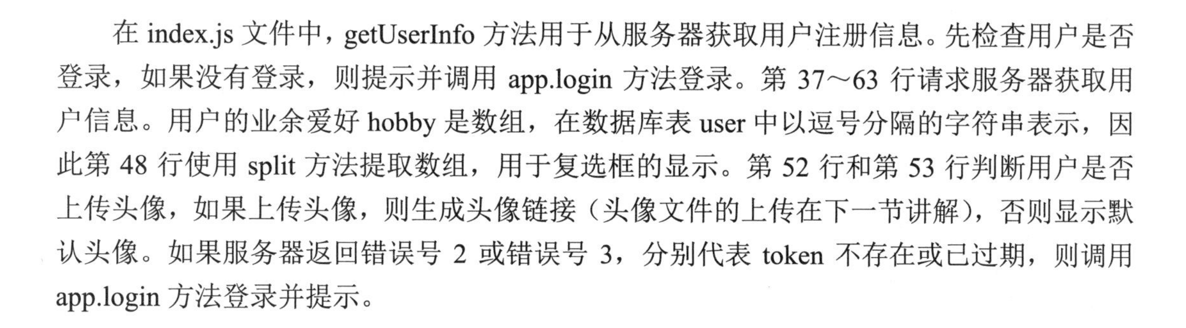

在 index.js 文件中，getUserInfo 方法用于从服务器获取用户注册信息。先检查用户是否登录，如果没有登录，则提示并调用 app.login 方法登录。第 37～63 行请求服务器获取用户信息。用户的业余爱好 hobby 是数组，在数据库表 user 中以逗号分隔的字符串表示，因此第 48 行使用 split 方法提取数组，用于复选框的显示。第 52 行和第 53 行判断用户是否上传头像，如果上传头像，则生成头像链接（头像文件的上传在下一节讲解），否则显示默认头像。如果服务器返回错误号 2 或错误号 3，分别代表 token 不存在或已过期，则调用 app.login 方法登录并提示。

4.6.3　处理文件上传

1．ThinkPHP 处理文件上传

通常需要先上传图片、视频等文件到服务器，然后在小程序中进行显示或播放。要实现文件上传，服务器需要处理文件上传的请求。

打开 app/controller/index.php 文件，增加一个 file 方法。

```
1       public function file()
2       {
3           if (request()->isPost()) {
4               //获取文件上传对象
5               $files = request()->file();
6               if (isset($files) && !empty($files)) {
7                   //文件上传后本地服务器的存储路径
8                   $savename = [];
9                   foreach ($files as $file) {
10                      $savename[] = Filesystem::disk('public')->putFile('files', $file);
11                      return json([
12                          'err_code' => 0,
13                          'file_path' => str_replace('\\', '/', $savename[0])
14                      ]);
15                  }
16              }
17          }
18      }
```

第 10 行负责将文件保存到 ThinkPHP 的默认文件夹中。第 11～14 行返回保存在服务器的完整文件名给客户端。

2．小程序实现文件上传

回到微信开发者工具，打开 index.js 文件，在 Page({})内部添加一个 changeAvatar 方法，输入以下代码：

```
1      changeAvatar: function (e) {
2        var that = this;
3        wx.chooseMedia({
4          count: 1, // 默认为9
5          mediaType: ['image'],
6          sizeType: ['original', 'compressed'], // 可以指定是原图还是压缩图，
   默认二者都有
7          sourceType: ['album', 'camera'], // 可以指定来源是相册还是相机，默认
   二者都有
8          success: function (res) {
9            // 返回选定图片的本地文件路径列表，tempFilePath 可以作为img标签的src
   属性显示图片
10           console.log(res)
11           wx.uploadFile({
12             url: settings.server_url + 'file',
13             filePath: res.tempFiles[0].tempFilePath,
14             name: 'file',
15             success: res => {
16               let result = JSON.parse(res.data)
17               that.avatar = result.file_path
18               that.setData({ avatar: settings.file_url + result.file_
   path })
19             }
20           })
21         }
22       })
23     }
```

这段代码调用了两个新的微信 API：wx.chooseMedia 和 wx.uploadFile，前者用于选择图片或视频，后者用于上传文件到服务器。

第 13 行中的 res.tempFiles[0].tempFilePath 代表选择的第一个文件的本地临时文件路径。由于第 4 行中的 count 设置为 1，因此只能选择一张图片。

第 11 行中的 wx.uploadFile 指定 url 是服务器编写的 index 控制器的 file 方法，也就是上面编写的 file 方法用于处理文件上传。在第 13 行指定了上传的文件路径，即选择图片文件的本地临时文件路径。wx.uploadFile 会将指定的文件上传到指定的 url，上传成功后在第 17 行将服务器的文件 url 指定给 Page({})内部的 avatar 变量，以便用户注册时提交给服务器保存。第 18 行设置头像文件的完整链接，刷新页面后，页面顶部头像将显示选择的图片文件。

wx.chooseMedia 用于拍摄，以及从相册中选择图片或视频。在传递的参数中，count 代表最多可选择文件的个数，基础库 2.25.0 之前版本最多支持 9 个文件，基础库 2.25.0 及以后版本最多支持 20 个文件。mediaType 代表文件类型为数组类型，数

组元素为“image”或“video”，分别代表图片或视频。sourceType 代表图片或视频选择的来源，也是数组类型，数组元素可以是“album”或“camera”，分别代表相册或相机拍摄。sizeType 代表是否压缩所选文件，基础库 2.25.0 之前版本仅在 mediaType 为 image 时有效，基础库 2.25.0 及以后版本对全量 mediaType 有效，也是数组类型，数组元素可以是“original”或“compressed”，分别代表原始大小或压缩。camera 仅在 sourceType 为 camera 时生效，取值可以是“back”或“front”，分别代表使用后置摄像头或前置摄像头。

4.6.4　处理用户注册

扫一扫

微课：用户注册-上传头像代码

1. ThinkPHP 处理用户注册

用户注册时需要将小程序提交的各种用户信息保存到数据库中。

打开 app/controller/index.php 文件，增加一个 register 方法，输入以下代码：

```
1        public function register()
2        {
3            if (request()->isPost()) {
4                $param = request()->param();
5                $token = $param['token'];
6                $result = $this->check_token($token);
7                if ($result['err_code'] > 0) {
8                    return json($result);
9                }
10               $openid = $result['openid'];
11               $user = DB::table('user')->where('openid', $openid)->find();
12               $count = 0;
13               $u = [
14               'openid' => $openid,
15                   'name' => $param['name'],
16                   'gender' => $param['gender'],
17                   'hobby' => $param['hobby'],
18                   'nationality' => $param['nationality'],
19                   'is_admin' => $param['is_admin'],
20                   'intro' => $param['intro']
21               ];
22               if (isset($param['avatar']))
23                   $u['avatar'] = $param['avatar'];
24               if ($user == null) {
25                   $count = DB::table('user')->insert($u);
26               } else {
27                   $count = DB::table('user')->save($u);
28               }
```

```
29              if ($count == 1) {
30                  return json([
31                      'err_code' => 0,
32                      'openid' => $openid
33                  ]);
34              } else {
35                  return json([
36                      'err_code' => 5,
37                      'err_msg' => $openid . '注册失败'
38                  ]);
39              }
40          } else {
41              return 'register';
42          }
43      }
```

第 11 行使用 openid 查找数据库表 user，第 13～21 行使用提交的数据生成用户对象 u（在 PHP 中称为数组），第 23 行判断是否在数据库表 user 中查找到用户，如果没有查找到，则运行第 24 行代码插入数据到数据库表 user，否则执行第 26 行代码修改数据到数据库表 user。

2．小程序实现用户注册

回到微信开发者工具，打开 index.js 文件，在 Page({})内部添加方法 register 和 submit，输入以下代码：

```
1       register: function (user) {
2         let token = app.globalData.token
3         if (!token) {
4           this.showMessage('还没有登录，请稍后重试')
5           app.login()
6           return
7         }
8         user.token = token
9         wx.request({
10          url: settings.server_url + 'register',
11          method: 'post',
12          data: user,
13          success: res => {
14            console.log(res.data)
15            if (res.data.err_code == 0) {
16              wx.showToast({
17                title: '注册成功',
18                duration: 1500,
19                icon: 'success'
20              })
21            } else {
22              this.showMessage('注册失败！')
```

```
23              //token 不存在或已过期
24              if (res.data.err_code == 2 || res.data.err_code == 3) {
25                app.login()
26              }
27            }
28          },
29          fail: err => {
30            console.error(err)
31          }
32        })
33      },
34      submit: function (e) {
35        console.log(e.detail.value)
36        let data = e.detail.value
37        if (data.name == '') {
38          this.showMessage('请输入姓名');
39          return
40        }
41        let user = data
42        user.hobby = data.hobby.join(',')
43        user.nationality = this.data.country[data.nationality]
44        if (this.avatar != '') user.avatar = this.avatar
45        this.register(user)
46      },
```

register 方法用于用户注册，即将传递过来的 user 参数提交到服务器进行保存。register 方法和 getUserInfo 方法类似，不再赘述。

submit 方法是用户单击“提交”按钮触发的事件处理方法。用户的业余爱好 hobby 是数组，在数据库表 user 中是以逗号分隔的字符串表示的，因此第 42 行将数组用逗号连接成字符串。第 44 行读取在文件上传处理中保存在 Page({})中的 avatar 变量。第 45 行将所有的用户注册信息通过 register 方法提交到服务器中保存。

至此，用户注册的服务器代码和小程序代码全部编写完成，运行后的截图如图 4.6.1 所示。

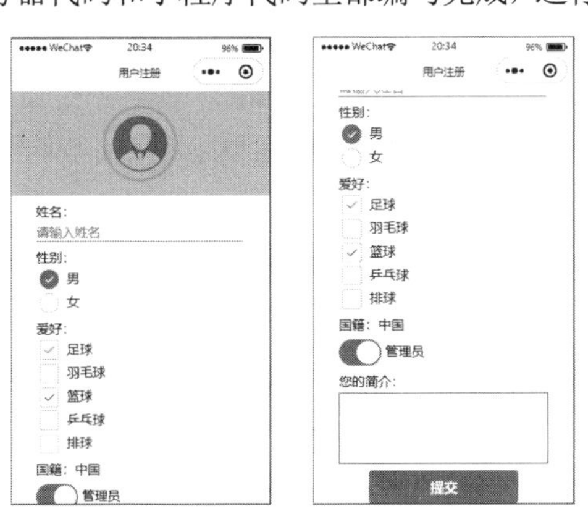

图 4.6.1　用户注册运行界面

单击头像可以选择头像图片文件，并将其上传到服务器。单击“提交”按钮可以提交所有的用户信息到服务器，并将其保存到数据库中。再次运行小程序时将自动读取头像和其他用户信息并显示。

项目小结

本项目内容比较复杂，如果没有系统学习过 PHP 编程语言和数据库的知识，难点在于 ThinkPHP 服务器的搭建和 MySQL 数据库的基本操作，使用 PHP 代码操作数据库是重点。

本项目在讲解微信小程序登录流程的基础上，完整地实现了用户注册过程，这几乎是任何一个商用微信小程序都需要的功能，因此实用性非常强。

结合本项目的案例，还可以学习 input、textarea、radio、checkbox、slider 和 switch 等常用组件的使用方法。

习题

一、判断题

1．wx.setStorage 将数据存储在本地缓存指定的 key 中，本地缓存的数据一直可用。（　　）

2．WXS 语法与 JavaScript 语法非常接近。（　　）

3．openid 是用户在当前小程序的唯一标识。（　　）

二、选择题

1．小程序使用 form 组件提交时，输入组件需要指定（　　）属性才能提交输入的值。

A．value　　B．name

C．type　　D．index

2．用户注册时将用户信息提交到后台服务器中保存，同时携带从服务器获取的（　　）来表明自己的身份，由服务器进行身份认证。

A．token　　B．openid

C．AppID　　D．AppSecret

3．通常在数据库中增加（　　）字段来与当前登录的用户进行绑定。

A．userid　　B．AppID

C．AppSecret　　D．openid

三、填空题

1．微信小程序登录时，开发者服务器把获得的________和小程序对应的 AppID、AppSecret 发送给微信接口服务。

2．输入多行文本时需要使用________________组件。

3．wx.chooseMedia 用于拍摄，以及从相册中选择图片或__________。

四、编程题

在现有的用户注册小程序基础上，增加用户的身份证号码、电话号码、地址、邮编等常见字段用于注册。

项目五　媒体播放器

任务一　播放音频

5.1.1　BackgroundAudioManager 对象

背景音频可以在微信小程序进入后台后继续播放，这是很多音乐播放器的常规做法。

要使用背景音频，需要调用 wx.getBackgroundAudioManager 方法得到全局唯一的 BackgroundAudioManager 对象，即背景音频管理器。

要使用背景音频，需要在 app.json 文件中配置：

```
"requiredBackgroundModes": ["audio"],
```

BackgroundAudioManager 对象的常用属性如表 5.1.1 所示。

表 5.1.1　BackgroundAudioManager 对象的常用属性

属性	含义
string src	音频的数据源，通常是网页链接，链接的对象必须是音频，不能是普通网页。目前支持的音频格式有 m4a、aac、mp3 和 wav。 需要特别注意的是：当设置了新的 src 属性时，会自动开始播放
number startTime	开始播放的位置（单位：秒）
string title	标题，用作原生音频播放器的标题（必填）
string epname	专辑名
string singer	歌手名
string coverImgUrl	封面图 URL，用作原生音频播放器的背景图
number duration	音频的长度（单位：秒），只读
number currentTime	音频的播放位置（单位：秒），只读
boolean paused	是否暂停，只读

BackgroundAudioManager 对象的常用方法如表 5.1.2 所示。

表 5.1.2　BackgroundAudioManager 对象的常用方法

方法	含义
play	播放音频
pause	暂停播放音频
seek(number currentTime)	跳转到指定位置
stop	停止播放音频
onCanplay(function listener)	监听背景音频进入可播放状态事件。通常在回调函数中调用 pause 方法来取消背景音频设置 src 属性后的自动播放
onError(function listener)	监听背景音频播放错误事件
onPlay(function listener)	监听背景音频播放事件
onPause(function listener)	监听背景音频暂停事件
onSeeking(function listener)	监听背景音频开始跳转事件
onSeeked(function listener)	监听背景音频完成跳转事件
onEnded(function listener)	监听背景音频自然播放结束事件。通常在回调函数中设置 src 属性是下一个音频的链接
onStop(function listener)	监听背景音频停止事件
onTimeUpdate(function listener)	监听背景音频播放进度更新事件，只有小程序在前台时才回调。通常在回调函数中刷新进度条，也可以更新音频的播放进度

下面通过一个简单的例子来演示背景音频的播放，详细代码演示见任务九。

打开微信开发者工具，使用测试账号新建一个微信小程序，不使用模板。打开 index.wxml 文件，删除原来的代码，输入以下代码：

```
1   <view class="container">
2     <button bindtap="play">播放</button>
3     <button bindtap="pause">暂停</button>
4   </view>
```

打开 index.js 文件，删除原来的代码，输入以下代码：

```
1    Page({
2      backAudio: null,
3      onReady: function() {
4        this.backAudio = wx.getBackgroundAudioManager()
5        this.backAudio.title = '罗刹海市'
6        this.backAudio.epname = '山歌寥哉'
7        this.backAudio.singer = '刀郎'
8        this.backAudio.coverImgUrl = 'https://img.zcool.cn/community/
   01f2595d920f13a801211d539c3893.jpg@1280w_1l_2o_100sh.jpg'
9        this.backAudio.src = 'https://music.163.com/song/media/outer/
   url?id=2063487880.mp3'
10     },
11     play: function(e) {
12       this.backAudio.play()
13     },
14     pause: function(e) {
```

```
15         this.backAudio.pause()
16       }
17     })
```

第 3 行的 onReady 事件是页面初次渲染成功后调用的回调函数。第 4 行调用 wx.getBackgroundAudioManager 方法得到全局唯一的 BackgroundAudioManager 对象。第 5～9 行设置了常用的属性，将播放歌曲“罗刹海市”。第 11 行的 play 方法是单击“播放”按钮的事件处理方法，调用 BackgroundAudioManager 对象的 play 方法播放。第 14 行的 pause 方法是单击“暂停”按钮的事件处理方法，调用 BackgroundAudioManager 对象的 pause 方法暂停播放。

打开 app.json 文件，在 windows 的配置后面添加以下配置：

```
"requiredBackgroundModes": ["audio"],
```

完整的代码如下：

```
1   {
2     "pages": [
3       "pages/index/index"
4     ],
5     "window": {
6       "backgroundTextStyle": "light",
7       "navigationBarBackgroundColor": "#fff",
8       "navigationBarTitleText": "Weixin",
9       "navigationBarTextStyle": "black"
10    },
11    "requiredBackgroundModes": ["audio"],
12    "style": "v2",
13    "sitemapLocation": "sitemap.json",
14    "lazyCodeLoading": "requiredComponents"
15  }
```

编译后会播放音频，单击“暂停”按钮暂停播放，单击“播放”按钮继续播放，如图 5.1.1 所示。

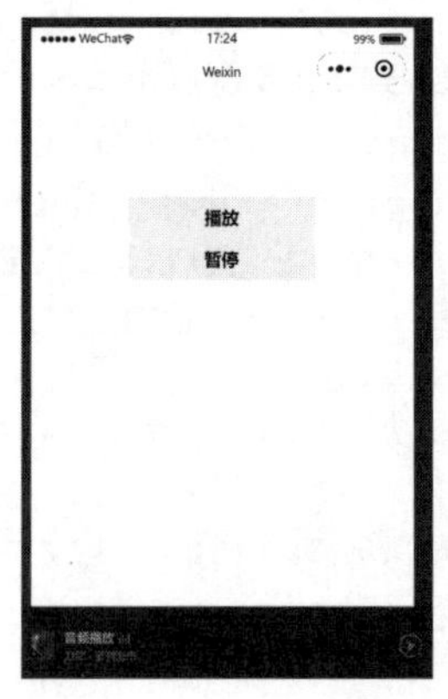

图 5.1.1 背景音频播放

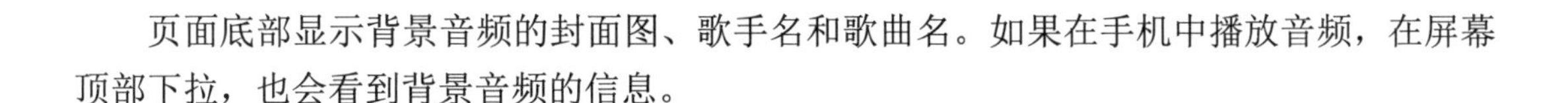

页面底部显示背景音频的封面图、歌手名和歌曲名。如果在手机中播放音频，在屏幕顶部下拉，也会看到背景音频的信息。

5.1.2 InnerAudioContext 对象

微信小程序还可以通过 wx.createInnerAudioContext 获取 InnerAudioContext 对象，用来播放音频，但在播放过程中可能被系统中断。InnerAudioContext 对象的属性、方法与 BackgroundAudioManager 对象接近。主要区别是 InnerAudioContext 对象比 BackgroundAudioManager 对象缺少 title、epname、singer、coverImgUrl 等属性，设置 src 属性后不会自动播放，必须调用 play 方法才会播放，也不需要在 app.json 文件中进行特殊设置。

打开 index.js 文件，为了便于代码对照，不删除原来的代码，注释不运行即可，修改后的代码如下：

```
1    Page({
2      // backAudio: null,
3      audio: null,
4      onReady: function() {
5        // this.backAudio = wx.getBackgroundAudioManager()
6        // this.backAudio.title = '罗刹海市'
7        // this.backAudio.epname = '山歌寥哉'
8        // this.backAudio.singer = '刀郎'
9        // this.backAudio.coverImgUrl = 'https://img.zcool.cn/community/
   01f2595d920f13a801211d539c3893.jpg@1280w_1l_2o_100sh.jpg'
10       // this.backAudio.src = 'https://music.163.com/song/media/outer/
   url?id=2063487880.mp3'
11       this.audio = wx.createInnerAudioContext()
12       this.audio.src = 'https://music.163.com/song/media/outer/url?id=
   2063487880.mp3'
13     },
14     play: function(e) {
15       // this.backAudio.play()
16       this.audio.play()
17     },
18     pause: function(e) {
19       // this.backAudio.pause()
20       this.audio.pause()
21     }
22   })
```

可以看到，这段代码与前面的代码非常接近。编译后不会播放音频，单击“播放”按钮进行播放，单击“暂停”按钮暂停播放，运行的页面与图 5.1.1 类似，只是底部没有背景音频部分。

任务二　播放视频

在微信小程序中播放视频可以使用 video 组件。video 组件的常用属性如表 5.2.1 所示。

表 5.2.1　video 组件的常用属性

属性	含义
string src	要播放视频的资源地址，支持网络路径、本地临时路径、云文件 ID
number duration	视频的时长（单位：秒）
boolean controls	是否显示默认播放控件（“播放”/“暂停”按钮、播放进度、时间）
Array danmu-list	弹幕列表
boolean danmu-btn	是否显示弹幕按钮，只在初始化时有效，不能动态变更
boolean enable-danmu	是否显示弹幕，只在初始化时有效，不能动态变更
boolean autoplay	是否自动播放
boolean loop	是否循环播放
boolean muted	是否静音播放
number initial-Time	指定视频初始播放位置（单位：秒）
string title	视频的标题，全屏时在顶部显示

放置 video 组件后，需要指定 id 属性，先在 js 代码中通过 wx.createVideoContext(string id)获取 VideoContext 实例，再调用该实例的方法来控制视频的播放、暂停、停止、跳转等。VideoContext 实例的常见方法如表 5.2.2 所示。

表 5.2.2　VideoContext 实例的常见方法

方法	含义
play	播放视频
pause	暂停播放视频
seek(number position)	跳转到指定位置
stop	停止播放视频
sendDanmu(Object data)	发送弹幕，data 参数指定 string text 和 string color，分别代表弹幕的文字和颜色

下面通过一个例子来演示如何使用 video 组件进行视频播放。

打开微信开发者工具，使用测试账号新建一个微信小程序，不使用模板。打开 index.wxml 文件，删除原来的代码，输入以下代码：

```
1   <view class="container">
2     <video id="myVideo" src="http://wxsnsdy.tc.qq.com/105/20210/snsdyvideodownload?filekey=30280201010421301f0201690402534804102ca905ce620b1241b726bc41dcff44e00204012882540400&bizid=1023&hy=SH&fileparam=302c020101042530230204136ffd93020457e3c4ff02024ef202031e8d7f02030f42400204045a320a0201000400" binderror="videoErrorCallback" danmu-list="{{danmuList}}" enable-danmu danmu-btn controls></video>
```

```
3      <input bindblur="inputBlur" placeholder="在此处输入弹幕内容" />
4      <button bindtap="sendDanmu" type="primary" formType="submit">发送
弹幕</button>
5      <button bindtap="play" type="primary">播放</button>
6      <button bindtap="pause" type="primary">暂停</button>
7    </view>
```

第 2 行使用 video 组件，指定了 id 属性，在 js 代码中使用这个属性生成 VideoContext 实例。第 3 行中的 bindblur 在 input 组件失去焦点时触发，当用户在 input 组件内输入弹幕后，单击“发送弹幕”按钮或其他组件时，会使 input 组件失去焦点。

打开 app.wxss 文件，将第 8 行中的 200rpx 修改为 20rpx。

打开 index.wxss 文件，删除原来的代码，输入以下代码：

```
1   input {
2     border: 2rpx #666 solid;
3     margin: 20rpx;
4     width: 80%;
5   }
6   button {
7     margin: 20rpx;
8   }
```

打开 index.js 文件，删除原来的代码，输入以下代码：

```
1   function getRandomColor() {
2     const rgb = []
3     for (let i = 0; i < 3; ++i) {
4       let color = Math.floor(Math.random() * 256).toString(16)
5       color = color.length == 1 ? '0' + color : color
6       rgb.push(color)
7     }
8     return '#' + rgb.join('')
9   }
10
11  Page({
12    onReady: function (res) {
13      this.videoContext = wx.createVideoContext('myVideo')
14    },
15    inputValue: '',
16    data: {
17      src: '',
18      danmuList:
19        [{
20          text: '第 1s 出现的弹幕',
21          color: '#ff0000',
```

```
22            time: 1
23          },
24          {
25            text: '第 3s 出现的弹幕',
26            color: '#ff00ff',
27            time: 3
28          }]
29      },
30      inputBlur: function (e) {
31        this.inputValue = e.detail.value
32      },
33      sendDanmu: function () {
34        this.videoContext.sendDanmu({
35          text: this.inputValue,
36          color: getRandomColor()
37        })
38      },
39      play: function () {
40        this.videoContext.play()
41      },
42      pause: function () {
43        this.videoContext.pause()
44      },
45      videoErrorCallback: function (e) {
46        console.log('视频错误信息:')
47        console.log(e.detail.errMsg)
48      }
49    })
```

在微信小程序中可以使用 HTML 中的#RRGGBB 颜色格式（RR 代表两位十六进制的红色数据，GG 代表绿色，BB 代表蓝色）。第 1～9 行中的 getRandomColor 方法用来生成这种格式的随机颜色。第 4 行中的 Math.random 表示返回 0～1 的随机浮点数（包含 0，但不包含 1），Math.floor 表示浮点数向下取整，toString(16)表示转换成十六进制的字符串，即得到 0～255 的整数再转换成十六进制数，第 5 行用于当第 4 行得到的字符串只有 1 位时，补充前导 0。第 6 行将字符串放入数组 rgb 中。第 8 行调用数组中的 join 方法，连接成一个字符串，补齐#后就是#RRGGBB 格式的颜色。

第 13 行得到 VideoContext 实例，其中的 id 参数必须与 wxml 文件中 video 的 id 属性值相同，否则得不到 VideoContext 实例。

编译后显示的页面如图 5.2.1 所示。

图 5.2.1　播放视频页面

任务三　使用轮播图

在专业网站中，在页面顶部经常可以看到轮播图，用于循环展示多张图片。在微信小程序中要实现这种效果非常简单，因为它提供了专门的滑块视图容器组件 swiper。swiper 组件使用简单，功能强大，不仅可以实现轮播图效果，还可以用于多页面滑动展示，在手机移动端得到了广泛应用。本任务主要完成轮播图功能，多页面滑动展示在任务七和任务九中都有应用。

swiper 组件是一个容器组件，内部只能放置 swiper-item 组件，代表各个滑块。swiper-item 组件内部可以放置各种常规组件，放置 image 组件就展现为轮播图。

swiper 组件的常用属性如表 5.3.1 所示。

表 5.3.1　swiper 组件的常用属性

属性	含义
boolean indicator-dots	是否显示面板指示点
color indicator-color	指示点颜色
color indicator-active-color	当前选中的指示点颜色
boolean autoplay	是否自动切换
number current	当前所在滑块的下标
number interval	自动切换时间间隔
number duration	滑块动画时长
boolean circular	是否采用衔接滑动
boolean vertical	滑动方向是否为纵向

swiper 组件还有一个常用的 bindchange 事件，当 current 改变时会触发 change 事件：event.detail={current,source}。该事件的演示见任务七和任务九。

下面举一个简单的例子来演示轮播图效果。

打开微信开发者工具，使用测试账号新建一个微信小程序，不使用模板。

在小程序根文件夹新建文件夹 images，使该文件夹与 pages 文件夹并列。任意放置多张图片到 images 文件夹中，为了显示效果，图片的大小尽量一致。

打开 index.wxml 文件，删除原来的代码，输入以下代码：

```
1    <swiper indicator-dots="#FFF" indicator-active-color="#FF4C91"
     autoplay="true" >
2      <swiper-item wx:for="{{['20230507.jpg', '20230508.jpg', '20230509.
     jpg', '20230510.jpg', '20230511.jpg', '20230512.jpg']}}" wx:key=
     "*this" >
3        <image mode="aspectFit" src="/images/{{item}}" />
4      </swiper-item>
5    </swiper>
```

注意：第 2 行使用列表渲染，wx:for 中的数组元素根据放置在 images 文件夹中的图片文件名和个数来确定，不能照搬代码。

打开 index.wxss 文件，删除原来的代码，输入以下代码：

```
1    swiper {
2      height: 500rpx;
3    }
4
5    image {
6      width: 100%;
7    }
```

不需要编写 js 代码。编译后显示的页面如图 5.3.1 所示。

图 5.3.1　轮播图页面

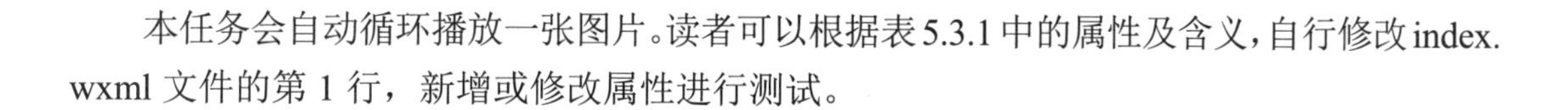

本任务会自动循环播放一张图片。读者可以根据表5.3.1中的属性及含义，自行修改index.wxml文件的第1行，新增或修改属性进行测试。

任务四　使用 tabBar

在手机应用中，底部的标签按钮很常见，一般将常用的几个功能做成标签按钮，单击按钮可以进入对应页面。在微信小程序中可以很容易地配置这种标签按钮：在app.json文件中配置tabBar，通过tabBar配置项指定标签按钮的颜色（选中状态或不选中状态可以分别设置）、标签栏的背景色、标签按钮的图片、标签按钮对应的页面等。

tabBar的属性如表5.4.1所示。

表 5.4.1　tabBar 的属性

属性	必填	默认值	描述
HexColor color	是		标签按钮上的文字的默认颜色，仅支持十六进制颜色
HexColor selectedColor	是		标签按钮上的文字被选中时的颜色，仅支持十六进制颜色
HexColor backgroundColor	是		标签按钮的背景色，仅支持十六进制颜色
Array list	是		标签按钮的列表，详见list属性说明，最少为2个标签按钮，最多为5个标签按钮
string borderStyle	否	black	标签按钮上边框的颜色，仅支持black和white
string position	否	bottom	标签按钮的位置，仅支持bottom和top

list属性用于定义标签按钮，如表5.4.2所示。

表 5.4.2　list 属性

属性	含义
string pagePath	标签按钮对应的页面路径
string text	标签按钮上的文字
string iconPath	标签按钮的图片
string selectedIconPath	当前选中的标签按钮图片

任务五中有一个实际的tabBar案例，代码如下：

```
1      "tabBar": {
2        "color": "#666666",
3        "selectedColor": "#1296db",
4        "backgroundColor": "#f6f6f6",
5        "borderStyle": "white",
6        "list": [
```

```
7          {
8            "pagePath": "pages/index/index",
9            "text": "音乐",
10           "iconPath": "images/music2.png",
11           "selectedIconPath": "images/music.png"
12         },
13         {
14           "pagePath": "pages/play_video/play_video",
15           "text": "视频",
16           "iconPath": "images/video2.png",
17           "selectedIconPath": "images/video.png"
18         },
19         {
20           "pagePath": "pages/admin/admin",
21           "text": "小编",
22           "iconPath": "images/admin2.png",
23           "selectedIconPath": "images/admin.png"
24         }
25       ]
26     },
```

读者可以参照表 5.4.1 和表 5.4.2 的描述理解 tabBar 的用法，在本任务中不再举例，示例在任务五中实现。

任务五　初始化媒体播放器项目

5.5.1　项目介绍

本项目要完成一个媒体播放器，可以播放音乐和视频，在后台可以添加需要播放音乐的栏目（专辑）、音乐及视频。

本项目包括三个主要功能，分别对应底部的三个按钮。

“音乐”按钮对应音乐播放，如图 5.5.1 所示。第一张图片是音乐播放器首页，显示所有音乐栏目及栏目下的音乐；单击某首音乐后，显示第二张图片，下面有五个功能按钮，分别是“播放模式”按钮（顺序播放、随机播放、单曲循环）、“上一首”按钮、“播放”/“暂停”按钮、“下一首”按钮和“播放列表”按钮；向右滑动屏幕显示歌词，对应第三张图片，会自动根据播放进度显示对应的歌词；第四张图片是播放列表，可以任选某首音乐播放。

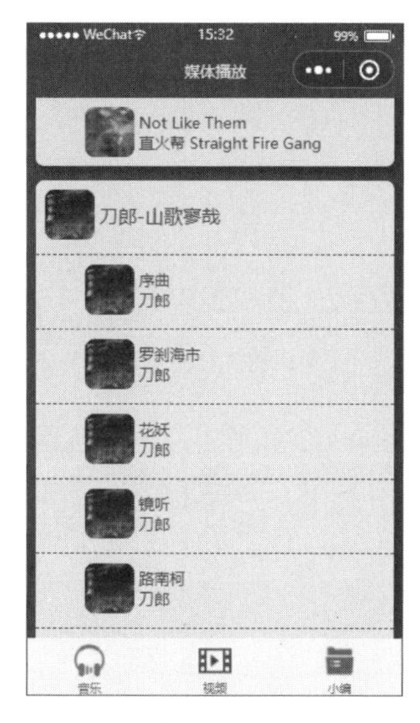

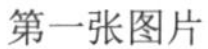
第一张图片

第二张图片

第四张图片

图 5.5.1 音乐播放器

单击底部的“视频”按钮进入视频页面，如图 5.5.2 所示。单击视频中间的“播放”/“暂停”按钮可以播放/暂停视频，向上滑动屏幕可以播放下一个视频。在视频播放页面可以输入弹幕，单击发送图片发送弹幕。

图 5.5.2 视频播放页面

单击“小编”按钮进入音乐、视频的编辑页面，如图 5.5.3 所示。该页面有三个标签页，分别是栏目、歌曲和视频，“标签”按钮位于页面顶部，可以单击或滑动屏幕进入。

图 5.5.3 “小编”页面

“栏目”页面显示音乐的栏目（或专辑），右下角有一个“增加”按钮，单击后显示如图 5.5.4 所示页面。或者单击某个栏目也可以进入如图 5.5.4 所示页面进行修改。先输入栏目名称，再选择一张图片作为栏目的图片，单击“保存”按钮，就会新增或修改栏目，并返回如图 5.5.3 所示页面。

图 5.5.4 “栏目”编辑页面

在如图 5.5.3 所示页面，单击顶部的“歌曲”标签按钮，或者向左滑动屏幕，可以进入“歌曲”编辑页面，如图 5.5.5 所示。单击右下角的“增加”按钮可以新增音乐，也可以单击某首音乐进行编辑，都会进入第二、三张图片显示的页面（因屏幕无法显示完整页面，故分成两张图片）。先分别输入歌名、歌手、歌曲 id，再选择图片和栏目，单击“保存”按钮即可新增或修改音乐，并回到如图 5.5.3 所示页面。一首音乐可以属于多个栏目，即栏目可以多选。

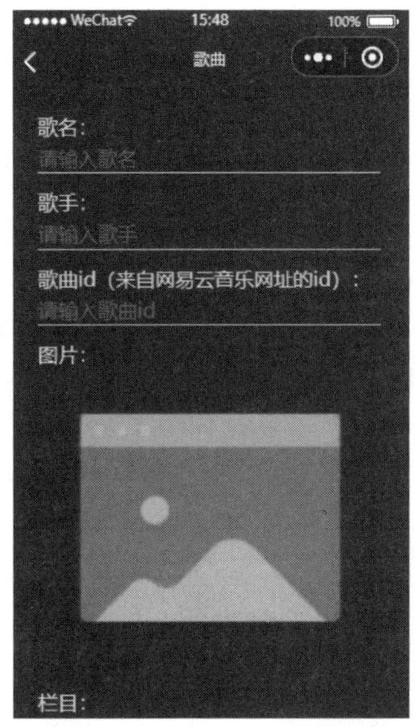

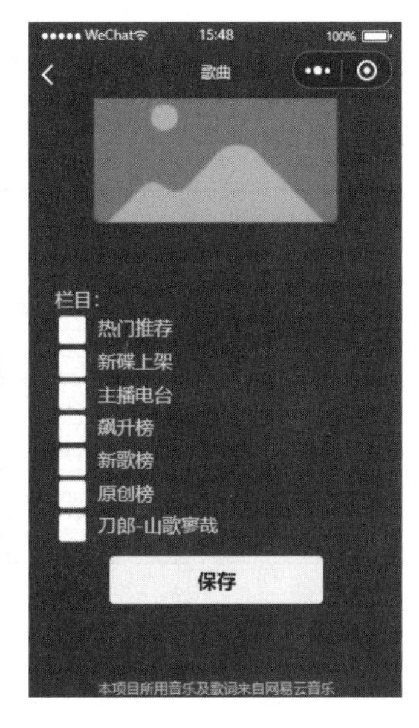

图 5.5.5　“歌曲”编辑页面

在如图 5.5.3 所示页面，单击顶部的“视频”标签按钮，或者向左滑动屏幕，可以进入“视频”编辑页面，如图 5.5.6 所示。该页面的操作方法与“歌曲”编辑页面类似，不再赘述。

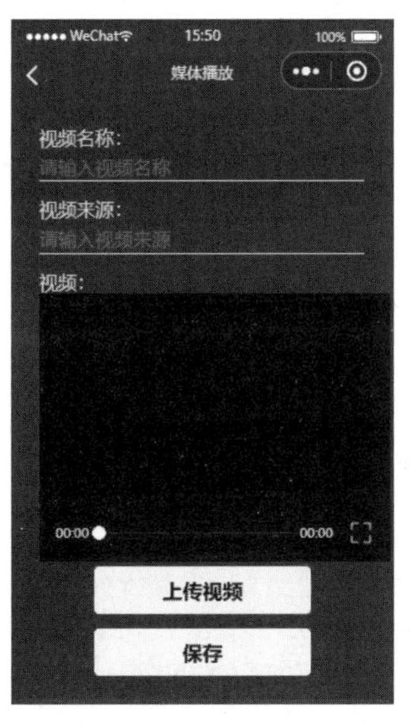

图 5.5.6　“视频”编辑页面

5.5.2 增加数据库表

本项目需要四张数据库表，分别是 category（栏目）、song（音乐）、video（视频）和 barrage（弹幕）。这四张表的定义如表 5.5.1 所示。

表 5.5.1 数据库表

表 category				
字段名	类型	长度	非空	备注
id	int	11	not null	主键、自动增长
name	varchar	50	not null	
img	varchar	255		
表 song				
字段名	类型	长度	非空	备注
id	int	11	not null	主键、自动增长
category	varchar	255	not null	
title	varchar	100	not null	
singer	varchar	100	not null	
src	varchar	255	not null	
img	varchar	255		
表 video				
字段名	类型	长度	非空	备注
id	int	11	not null	主键、自动增长
src	varchar	255	not null	
name	varchar	255		
表 barrage				
字段名	类型	长度	非空	备注
id	int	11	not null	主键、自动增长
vid	int	11	not null	
text	varchar	255	not null	
time	int	11	not null	

对应的创建表格的 sql 代码如下：

```
1      CREATE TABLE 'category'  (
2        'id' int(11) NOT NULL AUTO_INCREMENT,
3        'name' varchar(50) CHARACTER SET utf8 COLLATE utf8_unicode_ci NOT
   NULL,
4        'img' varchar(255) CHARACTER SET utf8 COLLATE utf8_unicode_ci
   NULL DEFAULT NULL,
5        PRIMARY KEY ('id') USING BTREE
6      ) ENGINE = MyISAM AUTO_INCREMENT = 8 CHARACTER SET = utf8 COLLATE =
   utf8_unicode_ci ROW_FORMAT = Dynamic;
7
```

```
8       CREATE TABLE 'song'  (
9         'id' int(11) NOT NULL AUTO_INCREMENT,
10        'category' varchar(255) CHARACTER SET utf8 COLLATE utf8_unicode_
   ci NOT NULL,
11        'title' varchar(100) CHARACTER SET utf8 COLLATE utf8_unicode_ci
   NOT NULL,
12        'singer' varchar(100) CHARACTER SET utf8 COLLATE utf8_unicode_ci
   NOT NULL,
13        'src' varchar(255) CHARACTER SET utf8 COLLATE utf8_unicode_ci NOT
   NULL,
14        'img' varchar(255) CHARACTER SET utf8 COLLATE utf8_unicode_ci
   NULL DEFAULT NULL,
15        PRIMARY KEY ('id') USING BTREE
16      ) ENGINE = MyISAM AUTO_INCREMENT = 35 CHARACTER SET = utf8 COLLATE =
   utf8_unicode_ci ROW_FORMAT = Dynamic;
17
18      CREATE TABLE 'video'  (
19        'id' int(11) NOT NULL AUTO_INCREMENT,
20        'src' varchar(255) CHARACTER SET utf8 COLLATE utf8_unicode_ci NOT
   NULL,
21        'name' varchar(255) CHARACTER SET utf8 COLLATE utf8_unicode_ci
   NULL DEFAULT NULL,
22        'source' varchar(255) CHARACTER SET utf8 COLLATE utf8_unicode_ci
   NULL DEFAULT NULL,
23        PRIMARY KEY ('id') USING BTREE
24      ) ENGINE = MyISAM AUTO_INCREMENT = 5 CHARACTER SET = utf8 COLLATE =
   utf8_unicode_ci ROW_FORMAT = Dynamic;
25
26      CREATE TABLE 'barrage'  (
27        'id' int(11) NOT NULL AUTO_INCREMENT,
28        'vid' int(11) NOT NULL,
29        'text' varchar(255) CHARACTER SET utf8 COLLATE utf8_unicode_ci
   NOT NULL,
30        'time' int(11) NOT NULL,
31        PRIMARY KEY ('id') USING BTREE
32      ) ENGINE = MyISAM AUTO_INCREMENT = 5 CHARACTER SET = utf8 COLLATE =
   utf8_unicode_ci ROW_FORMAT = Dynamic;
```

5.5.3　创建项目并配置 tabBar

打开微信开发者工具，使用自己的 AppID 新建项目，名称为 media，不使用模板。

复制资源文件夹 media 下的 images 文件夹到新建的 media 微信小程序文件夹中，使该文件夹与 pages 文件夹并列。

打开 app.json 文件，在"pages"的配置数组后面添加以下代码：

```
1      "pages": [
2        "pages/index/index",
3        "pages/play_video/play_video",
4        "pages/admin/admin",
5        "pages/category/category",
6        "pages/song/song",
7        "pages/play/play",
8        "pages/video/video"
9      ],
```

按 Ctrl+S 组合键保存，或者单击工具条上的“编译”按钮，微信开发者工具会在 pages 文件夹下新增六个页面。接下来配置 tabBar，完整的 app.json 文件代码如下：

```
27    {
28      "pages": [
29        "pages/index/index",
30        "pages/play_video/play_video",
31        "pages/admin/admin",
32        "pages/category/category",
33        "pages/song/song",
34        "pages/play/play",
35        "pages/video/video"
36      ],
37      "window": {
38        "backgroundTextStyle": "light",
39        "navigationBarBackgroundColor": "#21445a",
40        "navigationBarTitleText": "媒体播放",
41        "navigationBarTextStyle": "white"
42      },
43      "tabBar": {
44        "color": "#666666",
45        "selectedColor": "#1296db",
46        "backgroundColor": "#f6f6f6",
47        "borderStyle": "white",
48        "list": [
49          {
50            "pagePath": "pages/index/index",
51            "text": "音乐",
52            "iconPath": "images/music2.png",
53            "selectedIconPath": "images/music.png"
54          },
55          {
56            "pagePath": "pages/play_video/play_video",
57            "text": "视频",
58            "iconPath": "images/video2.png",
59            "selectedIconPath": "images/video.png"
60          },
```

```
61          {
62            "pagePath": "pages/admin/admin",
63            "text": "小编",
64            "iconPath": "images/admin2.png",
65            "selectedIconPath": "images/admin.png"
66          }
67        ]
68      },
69      "requiredBackgroundModes": [
70        "audio"
71      ],
72      "style": "v2",
73      "sitemapLocation": "sitemap.json"
74    }
```

第 43～68 行配置了 tabBar，第 69～71 行配置了背景音乐，本项目将使用背景音乐方式播放音乐。

5.5.4　添加工具模块

在小程序根文件夹新建文件夹 utils，在微信开发者工具 “资源管理器”底部的空白处右击，并在弹出的快捷菜单中选择“新建文件夹”命令。

在 utils 文件夹中新建文件 util.js，输入以下代码：

```
1     const ip = '***.***.***.***',
2       server_url = 'http://' + ip + '/media/',
3       file_url = 'http://' + ip + '/storage/'
4
5     function showMessage(str) {
6       wx.showToast({
7         title: str,
8         duration: 1500,
9         icon: 'none'
10      })
11    }
12
13    //media:image 或 video
14    function uploadMedia(media, success, fail) {
15      var that = this;
16      wx.chooseMedia({
17        count: 1, // 默认 9
18        mediaType: [media],
19        sizeType: ['original', 'compressed'], // 可以指定是原图还是压缩图，默认二者都有
20        sourceType: ['album', 'camera'], // 可以指定来源是相册还是相机，默认二
```

```
    者都有
21          success: function (res) {
              // 返回选定图片的本地文件路径列表，tempFilePath 可以作为 img 标签的 src
22  属性显示图片
23            console.log(res)
24            wx.uploadFile({
25              url: server_url + 'file',
26              filePath: res.tempFiles[0].tempFilePath,
27              name: 'file',
28              success: res => {
29                console.log(res.data)
30                let result = JSON.parse(res.data)
31                if (success) success(result)
32              },
33              fail: err => {
34                if (fail) fail(err)
35              }
36            })
37          }
38        })
39      }
40      module.exports = {
41        server_url,
42        file_url,
43        showMessage,
44        uploadMedia
45      }
```

utils/util.js 模块用来配置服务器及文件上传的 URL，并提供了 showMessage 和 uploadMedia 两种共用方法。

在第 1 行需要输入自己电脑的 IP 地址。要查看电脑的 IP 地址，先输入 win+R，再输入 cmd 进入 Windows 的控制台，最后在控制台输入 ipconfig 即可。

第 2 行是服务器的 URL。用户注册需要上传头像文件到服务器，第 3 行是上传文件的 URL。

第 14 行的 uploadMedia 方法用于上传音乐或视频。

至此，媒体播放器的初始化工作完成。

任务六　使用 ThinkPHP 实现数据库的基本操作

扫一扫

微课：媒体播放-数据库代码

在 app/controller 文件夹下新建文件 Media.php，新建控制器类 Media，并添加有关栏目和音乐的方法，完整代码如下：

```
1	<?php
2	namespace app\controller;
3	
4	use app\BaseController;
5	use think\facade\Db;
6	use think\facade\Filesystem;
7	
8	
9	lass Media extends BaseController
10	{
11	    public function addCategory()
12	    {
13	        if (request()->isPost()) {
14	            $param = request()->param();
15	            $data = $param['data'];
16	            $count = DB::table('category')->insert($data);
17	            if ($count == 1) {
18	                return json([
19	                    'err_code' => 0
20	                ]);
21	            } else {
22	                return json([
23	                    'err_code' => 1,
24	                    'err_msg' => '增加栏目失败'
25	                ]);
26	            }
27	        } else {
28	            return 'addCategory';
29	        }
30	    }
31	
32	    public function updateCategory()
33	    {
34	        if (request()->isPost()) {
35	            $param = request()->param();
36	            $data = $param['data'];
37	            $count = DB::table('category')->update($data);
38	            if ($count == 1) {
39	                return json([
40	                    'err_code' => 0
41	                ]);
42	            } else {
43	                return json([
44	                    'err_code' => 2,
45	                    'err_msg' => '更新栏目失败'
46	                ]);
```

```
47                }
48            } else {
49                return 'updateCategory';
50            }
51        }
52
53        public function getCategory()
54        {
55            $param = request()->param();
56            $id = $param['id'];
57            $category = DB::table('category')->where('id', $id)->find();
58            if ($category == null) {
59                return json([
60                    'err_code' => 3,
61                    'err_msg' => $id . '栏目不存在'
62                ]);
63            } else {
64                return json([
65                    'err_code' => 0,
66                    'category' => $category
67                ]);
68            }
69        }
70
71        public function getAllCategories()
72        {
73            $categories = DB::table('category')->select()->toArray();
74            return json([
75                'err_code' => 0,
76                'categories' => $categories
77            ]);
78        }
79
80        public function addSong()
81        {
82            if (request()->isPost()) {
83                $param = request()->param();
84                $data = $param['data'];
85                $count = DB::table('song')->insert($data);
86                if ($count == 1) {
87                    return json([
88                        'err_code' => 0
89                    ]);
90                } else {
91                    return json([
92                        'err_code' => 1,
```

```
93                      'err_msg' => '增加歌曲失败'
94                  ]);
95              }
96          } else {
97              return 'addSong';
98          }
99      }
100
101     public function updateSong()
102     {
103         if (request()->isPost()) {
104             $param = request()->param();
105             $data = $param['data'];
106             $count = DB::table('song')->update($data);
107             if ($count == 1) {
108                 return json([
109                     'err_code' => 0
110                 ]);
111             } else {
112                 return json([
113                     'err_code' => 2,
114                     'err_msg' => '更新歌曲失败'
115                 ]);
116             }
117         } else {
118             return 'updateSong';
119         }
120     }
121
122     public function getSong()
123     {
124         $param = request()->param();
125         $id = $param['id'];
126         $song = DB::table('song')->where('id', $id)->find();
127         if ($song == null) {
128             return json([
129                 'err_code' => 3,
130                 'err_msg' => $id . '歌曲不存在'
131             ]);
132         } else {
133             return json([
134                 'err_code' => 0,
135                 'song' => $song
136             ]);
137         }
138     }
```

```
139
140        public function getSongsByCategory()
141        {
142            $param = request()->param();
143            $id = $param['id'];
144            //song.category 是使用逗号分隔的栏目 id，一首歌曲可以属于多个栏目
145            $songs = Db::table('song')->where('find_in_set(:id,
category)', ['id' => $id])->select();
146            return json([
147                'err_code' => 0,
148                'songs' => $songs
149            ]);
150        }
151
152        public function addVideo()
153        {
154            if (request()->isPost()) {
155                $param = request()->param();
156                $data = $param['data'];
157                $count = DB::table('video')->insert($data);
158                if ($count == 1) {
159                    return json([
160                        'err_code' => 0
161                    ]);
162                } else {
163                    return json([
164                        'err_code' => 1,
165                        'err_msg' => '增加视频失败'
166                    ]);
167                }
168            } else {
169                return 'addVideo';
170            }
171        }
172
173        public function updateVideo()
174        {
175            if (request()->isPost()) {
176                $param = request()->param();
177                $data = $param['data'];
178                $count = DB::table('video')->update($data);
179                if ($count == 1) {
180                    return json([
181                        'err_code' => 0
182                    ]);
183                } else {
```

```
184                 return json([
185                     'err_code' => 2,
186                     'err_msg' => '更新视频失败'
187                 ]);
188             }
189         } else {
190             return 'updateVideo';
191         }
192     }
193 
194     public function getVideo()
195     {
196         $param = request()->param();
197         $id = $param['id'];
198         $video = DB::table('video')->where('id', $id)->find();
199         if ($video == null) {
200             return json([
201                 'err_code' => 3,
202                 'err_msg' => $id . '视频不存在'
203             ]);
204         } else {
205             return json([
206                 'err_code' => 0,
207                 'video' => $video
208             ]);
209         }
210     }
211 
212     public function getAllVideos()
213     {
214         $videos = DB::table('video')->select()->toArray();
215         return json([
216             'err_code' => 0,
217             'videos' => $videos
218         ]);
219     }
220 
221     public function addBarrage()
222     {
223         if (request()->isPost()) {
224             $param = request()->param();
225             $data = $param['data'];
226             $count = DB::table('barrage')->insert($data);
227             if ($count == 1) {
228                 return json([
229                     'err_code' => 0
```

```
230                 ]);
231             } else {
232                 return json([
233                     'err_code' => 1,
234                     'err_msg' => '增加弹幕失败'
235                 ]);
236             }
237         } else {
238             return 'addBarrage';
239         }
240     }
241 
242     public function getBarragesByVideoId()
243     {
244         $param = request()->param();
245         $id = $param['id'];
246         $barrages = Db::table('barrage')->where('vid', $id)->select();
247         return json([
248             'err_code' => 0,
249             'barrages' => $barrages
250         ]);
251     }
252 
253     public function file()
254     {
255         if (request()->isPost()) {
256             //获取文件上传对象
257             $files = request()->file();
258             if (isset($files) && !empty($files)) {
259                 //文件上传到本地服务器后的存储路径
260                 $savename = [];
261                 foreach ($files as $file) {
262                     $savename[] = Filesystem::disk('public')->putFile('files', $file);
263                     return json([
264                         'err_code' => 0,
265                         'file_path' => str_replace('\\', '/', $savename[0])
266                     ]);
267                 }
268             }
269         }
270     }
271 }
```

第 253 行的 file 方法与项目四的文件上传方法相同，其他方法都是基本的 ThinkPHP 数据库操作，不再一一讲述，仅对第 145 行做说明。

一首音乐可以属于多个栏目，音乐的 category 字段是由使用逗号分隔的栏目 id 组成的字符串。MySQL 数据库使用一个特殊的方法 find_in_set 来处理这种情况。

find_in_set 的官方解释是“假如字符串 str 在由 *N* 个子链组成的字符串列表 strlist 中，则返回值的范围在 1 到 *N* 之间”。例如，“select find_in_set('b', 'a,b,c,d ');”返回 2，“select find_in_set('g', 'a,b,c,d ');”返回 0。

“where('find_in_set(:id,category)', ['id' => $id])”的意思是如果指定的栏目 id 在某首音乐的 category 字段中，则返回 id 在 category 中的位置，否则返回 0。因此“$songs = Db::table('song')->where('find_in_set(:id,category)', ['id' => $id])->select();”整条语句的意思是查询出音乐的 category 字段中包含指定栏目 id 的所有音乐。

添加以上代码后，服务器可以实现栏目、音乐、视频、弹幕的增加、修改和基本的查询。

任务七　编辑栏目及音乐

扫一扫

微课：媒体播放-标签页切换代码

首先进行歌曲的编辑，因为没有歌曲就无法播放。

5.7.1　实现标签页切换

从媒体播放器的介绍知道，“小编”页面有三个标签页，分别是栏目、歌曲和视频。单击顶部的标签或左右滑动屏幕可以进行页面切换。要实现这些功能，需要标签栏和滑块视图容器 swiper。

打开 admin.wxml 文件，删除原来的代码，输入以下代码：

```
1      <view class="tab">
2        <view class="tab-item {{tab==0?'active':''}}" bindtap="tabTap" 
   id="0">栏目</view>
3        <view class="tab-item {{tab==1?'active':''}}" bindtap="tabTap" 
   id="1">歌曲</view>
4        <view class="tab-item {{tab==2?'active':''}}" bindtap="tabTap" 
   id="2">视频</view>
5      </view>
6      <view class="content">
7        <swiper current="{{item}}" bindchange="itemTap">
8          <swiper-item>
9            栏目
10         </swiper-item>
11         <swiper-item>
12           歌曲
```

```
13          </swiper-item>
14          <swiper-item>
15            视频
16          </swiper-item>
17        </swiper>
18      </view>
19      <image class="add" src="/images/add.png" bindtap="add"></image>
```

第 1～5 行实现标签栏，第 7～17 行中的 swiper 有三个滑块，分别对应栏目、歌曲和视频三个页面。

打开 admin.wxss 文件，输入以下代码：

```
1     page {
2       display: flex;
3       flex-direction: column;
4       background: #21445a;
5       color: #e9eaec;
6       height: 100%;
7     }
8
9     .content{
10      padding: 50rpx;
11    }
12
13    .content>swiper {
14      height: 95vh;
15    }
16
17    .tab {
18      display: flex;
19    }
20
21    .tab-item {
22      flex: 1;
23      font-size: 10pt;
24      text-align: center;
25      line-height: 72rpx;
26      border-bottom: 5rpx solid #e9eaec;
27    }
28
29    .tab-item.active {
30      border-top-left-radius: 20rpx;
31      border-top-right-radius: 20rpx;
32      color: #21445a;
33      background: #e9eaec;
34      font-weight: bold;
35      font-size: 12pt;
```

```
36    }
37
38    .add {
39      width: 80rpx;
40      height: 80rpx;
41      position: fixed;
42      bottom: 12rpx;
43      right: 12rpx;
44      border-radius: 50%;
45      background: #e9eaec;
46    }
```

第 29 行中的.tab-item.active 样式代表激活的标签样式。

打开 admin.js 文件，删除原来的代码，输入以下代码：

```
1   Page({
2
3     data: {
4       tab: 0,
5       item: 0,
6     },
7
8     onLoad(options) {
9     },
10
11    tabTap: function (e) {
12      this.setData({
13        item: Number(e.currentTarget.id)
14      })
15    },
16
17    itemTap: function (e) {
18      this.setData({
19        tab: e.detail.current,
20        item: e.detail.current
21      })
22    },
23  })
```

单击“标签”按钮会触发第 11 行中的 tabTap 事件，该事件设置了 item。刷新页面后修改 swiper 的 current，从而实现 swiper 的切换，还触发第 17 行中的 itemTap 事件，从而修改 tab 值，因此页面刷新后标签页会激活对应 tab 的显示。swiper 滑动时，也会触发 itemTap 事件。

通过标签栏和 swiper 的配合，实现标签页的切换。

5.7.2 编辑栏目和音乐

打开 admin.wxml 文件，将光标移动到“栏目”一行，按 Ctrl+/注释一行，并在后面输入以下代码：

```
1         <swiper-item>
2           <!-- 栏目 -->
3           <view class="list-empty" wx:if="{{categories.length === 0}}">
没有栏目</view>
4           <scroll-view scroll-y="true" class="scrolly">
5             <view class="list-item" wx:for="{{categories}}" wx:key="id"
bindtap="category" data-id="{{item.id}}">
6               <image class="list-img" src="{{item.imgUrl}}" />
7               <view class="list-info">{{item.name}}</view>
8               <image class="arrow" src="/images/arrow.png"></image>
9             </view>
10          </scroll-view>
11        </swiper-item>
```

将光标移动到“歌曲”一行，按 Ctrl+/注释一行，并在后面输入以下代码：

```
1         <swiper-item>
2           <!—歌曲 -->
3           <picker bindchange="changeIndex" value="{{index}}" range=
"{{categories}}" range-key="name">
4             <view>
5               音乐栏目：{{categories[index].name}}
6             </view>
7           </picker>
8           <view class="list-empty" wx:if="{{songs.length === 0}}">没有歌曲
</view>
9           <scroll-view scroll-y="true" class="scrolly">
10            <view class="list-item" wx:for="{{songs}}" wx:key="id"
bindtap="song" data-id="{{item.id}}">
11              <image class="list-img" src="{{item.imgUrl}}" />
12              <view class="list-info">
13                <view>{{item.title}}</view>
14                <view>{{item.singer}}</view>
15              </view>
16              <image class="arrow" src="/images/arrow.png"></image>
17            </view>
18          </scroll-view>
19        </swiper-item>
```

第 9 行中的样式.scrolly 是全局样式，需要添加在 app.wxss 文件中。打开 app.wxss 文件，插入以下代码：

```
1	.scrolly {
2	  height: 95%;
3	}
```

打开 admin.wxss 文件，在后面插入以下代码：

```
47	.list-item {
48	  display: flex;
49	  align-items: center;
50	  border: 1rpx solid #e9eaec;
51	  margin: 20rpx;
52	  border-radius: 10rpx;
53	  box-shadow: 10rpx 10rpx 10rpx 5rpx rgba(0, 0, 0, 0.4);
54	  height: 130rpx;
55	}
56	
57	.list-empty {
58	  margin: 80rpx, auto;
59	  padding: 50rpx;
60	  text-align: center;
61	}
62	
63	.list-img {
64	  width: 80rpx;
65	  height: 80rpx;
66	  margin-left: 15rpx;
67	  border-radius: 8rpx;
68	  border: 1px solid #21445a;
69	}
70	
71	.list-info {
72	  flex: 1;
73	  font-size: 10pt;
74	  line-height: 38rpx;
75	  margin-left: 20rpx;
76	  padding-bottom: 8rpx;
77	}
78	
79	.arrow {
80	  width: 50rpx;
81	  height: 50rpx;
82	  margin-right: 15rpx;
83	}
```

打开 admin.js 文件，在前面插入以下代码：

```
1	const util = require('../../utils/util.js')
2	
3	let that
```

在 data{}中插入以下代码：

```
1        categories: [],
2        songs: [],
3        index: 0,
```

在 Page({})中插入以下代码：

```
1      getAllCategories: function () {
2        wx.request({
3          url: util.server_url + 'getAllCategories',
4          success: res => {
5            console.log(res.data)
6            if (res.data.err_code == 0) {
7              res.data.categories.forEach(c => {
8                c.imgUrl = util.file_url + c.img
9              })
10             that.setData({
11               categories: res.data.categories
12             })
13             that.setCategoryId()
14             that.getSongsByCategory(this.categoryId)
15           }
16         },
17         fail: err => {
18           console.error(err)
19         }
20       })
21     },
22
23     categoryId: 0,
24     getSongsByCategory: function (id) {
25       if (id == 0) return
26       wx.request({
27         url: util.server_url + 'getSongsByCategory?id=' + id,
28         success: res => {
29           console.log(res.data)
30           if (res.data.err_code == 0) {
31             res.data.songs.forEach(s => {
32               s.imgUrl = util.file_url + s.img
33             })
34             that.setData({
35               songs: res.data.songs
36             })
37           }
38         },
39         fail: err => {
```

```
40          console.error(err)
41        }
42      })
43    },
44
45    setCategoryId: function (e) {
46      if (this.data.categories.length > 0)
47        this.categoryId = this.data.categories[this.data.index].id
48    },
49
50    onLoad(options) {
51      that = this
52      this.getAllCategories()
53    },
54
55    add: function (e) {
56      switch (this.data.item) {
57        case 0:
58          wx.navigateTo({
59            url: '/pages/category/category',
60            events: {
61              refresh: function (data) {
62                that.getAllCategories()
63              }
64            }
65          })
66          break
67        case 1:
68          wx.navigateTo({
69            url: '/pages/song/song',
70            events: {
71              refresh: function (data) {
72                that.getSongsByCategory(that.categoryId)
73              }
74            }
75          })
76          break
77        case 2:
78          break
79      }
80    },
81
82    category: function (e) {
83      wx.navigateTo({
84        url: '/pages/category/category?id=' +
85          e.currentTarget.dataset.id,
```

```
86          events: {
87            refresh: function (data) {
88              that.getAllCategories()
89            }
90          },
91        })
92      },
93
94      changeIndex: function (e) {
95        this.setData({ index: e.detail.value })
96        this.setCategoryId()
97        this.getSongsByCategory(this.categoryId)
98      },
99
100     song: function (e) {
101       wx.navigateTo({
102         url: '/pages/song/song?id=' +
103           e.currentTarget.dataset.id,
104         events: {
105           refresh: function (data) {
106             that.getSongsByCategory(that.categoryId)
107           }
108         },
109       })
110     },
```

第 1 行中的 getAllCategories 用于从服务器读取所有栏目。第 7～9 行使用 forEach 方法为读取的所有栏目添加一个 imgUrl 属性，用于保存栏目的图片链接。

第 24 行中的 getSongsByCategory 方法根据栏目 id 查找该栏目的所有音乐。第 32 行为所有音乐增加一个 imgUrl 属性，用于保存音乐的图片链接。

add 方法是单击右下角的“增加”按钮触发的事件处理方法。this.data.item 代表当前激活的标签页编号，0 代表栏目，1 代表歌曲，2 代表视频。当前标签页是栏目时，单击“增加”按钮则新增栏目；当前标签页是歌曲时，单击“增加”按钮则新增歌曲。

第 82 行中的 category 方法和第 100 行中的 song 方法是单击某个栏目或某首歌曲时，跳转到新页面去修改，相关代码在下一节介绍。

第 94 行中的 changeIndex 方法是单击栏目 picker 组件触发的事件处理方法，方法内部刷新栏目 index，并刷新栏目 id 和该栏目下的所有歌曲。

5.7.3 跳转栏目或音乐页面

微信小程序的页面分为 tabBar 页面和非 tabBar 页面，在 app.json 文件的 tabBar 属性中配置的页面都是 tabBar 页面，否则都是非 tabBar 页面。不同页面之间可以跳转，但跳转的方法不同，如表 5.7.1 所示。

表 5.7.1　不同页面之间的跳转方法

方法	说明
wx.switchTab	跳转到 tabBar 页面，并关闭其他所有非 tabBar 页面
wx.reLaunch	关闭所有页面，打开应用内的某个页面，tabBar 页面或非 tabBar 页面都可以
wx.redirectTo	关闭当前页面，跳转到非 tabBar 页面，不允许跳转到 tabBar 页面
wx.navigateTo	保留当前页面，跳转到非 tabBar 页面，不允许跳转到 tabBar 页面。使用 wx.navigateBack 可以返回原页面。小程序中的页面栈最多为十层

前三个方法所带的参数都相同，url 代表跳转的页面路径，success、fail、complete 分别是调用成功、调用失败、调用结束的回调函数。

下面重点介绍 wx.navigateTo 方法和 wx.navigateBack 方法。

调用 wx.navigateTo 方法跳转的页面与当前页面之间可以通信，互传数据，为多页面之间的互动带来很大的便利。

通信是通过调用 wx.navigateTo 方法的 events 参数进行的，该参数是页面之间通信的接口，用于监听被打开页面发送到当前页面的数据。

扫一扫

微课：媒体播放-页面跳转代码

当前页面向跳转页面传递数据有两种方法，一种是通过 url 带参数的方式传递数据，一种是在 success 回调函数中直接发送消息。如下列示例代码（这些代码仅是示例，不需要添加到本项目中）：

```
1   wx.navigateTo({
2     url: '/pages/test/test?id=1',
3     events: {
4       // 为指定事件添加一个监听器，获取被打开页面传递到当前页面的数据
5       backEvent1: function(data) {
6         console.log(data)
7       },
8       backEvent2: function(data) {
9         console.log(data)
10      }
11    },
12    success: function(res) {
13      // 通过 eventChannel 向被打开页面传递数据
14      res.eventChannel.emit('goEvent', { data: 'test' })
15    }
16  })
17  //  /pages/test/test.js
18  Page({
19    onLoad: function(option){
20      console.log(option.query)
21      const eventChannel = this.getOpenerEventChannel()
22      eventChannel.emit('backEvent1', {data: 'test'});
23      eventChannel.emit('backEvent2', {data: 'test'});
```

```
24        // 监听 goEvent 事件，获取上一页面通过 eventChannel 传递到当前页面的数据
25        eventChannel.on('goEvent', function(data) {
26          console.log(data)
27        })
28      }
29    })
```

第 2 行通过 url 带参数的方式传递数据，第 20 行通过 option.query 获取参数。在第 12～14 行，success 回调函数通过 res.eventChannel（事件通道）向跳转页面发送消息，跳转页面在第 25 行通过 eventChannel.on 接收。

第 22 行和第 23 行，跳转页面通过 eventChannel 向原来的页面发送了两个消息，分别在第 5 行和第 8 行接收。

在上一节最后的代码中，第 58 行跳转到 category 页面，第 61～63 行用于接收 category 页面的消息来刷新栏目。同样地，第 68 行跳转到 song 页面，第 71～73 行用于接收 song 页面的消息来刷新音乐。第 83 行也跳转到 category 页面，但是通过 url 带栏目 id 参数的方式，以方便 category 页面显示该栏目来进行修改。同样地，第 101 行跳转到 song 页面，通过 url 带音乐 id 参数的方式。

扫一扫

微课：媒体播放-栏目歌曲页面代码

下面完成 category 页面。

打开 category.wxml 文件，删除原来的代码，输入以下代码：

```
1    <view class="container">
2      <form bindsubmit="submit">
3        <view>
4          <text>栏目名称：</text>
5          <input name="name" placeholder="请输入栏目名称" value=
     "{{category.name}}" />
6        </view>
7        <view>
8          <text>图片：</text>
9          <image src="{{img}}" mode="aspectFit" bindtap="changeImg"></image>
10       </view>
11       <button form-type="submit">保存</button>
12     </form>
13   </view>
```

打开 category.wxss 文件，输入以下代码：

```
1    page {
2      background: #21445a;
3      color: #e9eaec;
4      height: 100%;
5    }
6
7    .container {
8      padding: 50rpx;
```

```
9	}
10	
11	view {
12	  margin-bottom: 30rpx;
13	}
14	
15	input {
16	  width: 650rpx;
17	  margin-top: 10rpx;
18	  border-bottom: 2rpx solid #e9eaec;
19	}
20	
21	image {
22	  width: 650rpx;
23	  border-radius: 20rpx;
24	}
```

打开 category.js 文件，输入以下代码：

```
1	const util = require('../../utils/util.js')
2	
3	let that
4	Page({
5	
6	  data: {
7	    category: {},
8	    img: '/images/no_image.png'
9	  },
10	
11	  getCategoryById: function (id) {
12	    wx.request({
13	      url: util.server_url + 'getCategory',
14	      method: 'post',
15	      data: { id: id },
16	      success: res => {
17	        console.log(res.data)
18	        if (res.data.err_code == 0) {
19	          that.img = res.data.category.img
20	          that.setData({
21	            category: res.data.category,
22	            img: util.file_url + that.img
23	          })
24	        } else if (res.data.err_code == 3) {
25	          util.showMessage('栏目不存在')
26	        }
27	      },
28	      fail: err => {
```

```
29            console.error(err)
30          }
31        })
32      },
33
34      id: '',
35      img: '',
36      eventChannel: null,
37      onLoad: function (options) {
38        that = this
39        this.eventChannel = this.getOpenerEventChannel()
40        if (options.id) {
41          this.id = options.id
42          this.getCategoryById(options.id)
43        }
44      },
45
46      changeImg: function (e) {
47        util.uploadMedia('image',
48          res => {
49            that.img = res.file_path
50            that.setData({ img: util.file_url + res.file_path })
51          },
52          err => {
53            console.error(err)
54          })
55      },
56
57      submit: function (e) {
58        var data = e.detail.value
59        data.img = this.img
60        if (this.id) {
61          data.id = this.id
62          wx.request({
63            url: util.server_url + 'updateCategory',
64            method: 'post',
65            data: { data: data },
66            success: res => {
67              console.log(res.data)
68              if (res.data.err_code == 0) {
69                console.log(that.eventChannel)
70                that.eventChannel.emit('refresh', { data: true })
71                wx.navigateBack()
72              } else if (res.data.err_code == 2) {
73                util.showMessage('更新栏目失败')
74              }
```

```
75              },
76              fail: err => {
77                console.error(err)
78              }
79            })
80          } else {
81            wx.request({
82              url: util.server_url + 'addCategory',
83              method: 'post',
84              data: { data: data },
85              success: res => {
86                console.log(res.data)
87                if (res.data.err_code == 0) {
88                  console.log(res)
89                  that.eventChannel.emit('refresh', { data: true })
90                  wx.navigateBack()
91                } else if (res.data.err_code == 1) {
92                  util.showMessage('增加栏目失败')
93                }
94              },
95              fail: err => {
96                console.error(err)
97              }
98            })
99          }
100       },
101
102     })
```

第41行通过url带的参数读取栏目id。第70行和第89行通过eventChannel向admin页面发送refresh消息，使admin重新读取栏目，并通过wx.navigateBack方法返回admin页面。

新增栏目和修改栏目共用一个category页面，通过是否指定id来区分，如果指定了id，则修改栏目，如第60行；否则新增栏目，如第80行。

第46行中的changeImg用于上传栏目图片，调用util.uploadMedia方法进行上传。

至此category页面完成，接下来完成song页面。

打开song.wxml文件，删除原来的代码，输入以下代码：

```
1     <view class="container">
2       <form bindsubmit="submit">
3         <view>
4           <text>歌名：</text>
5           <input name="title" placeholder="请输入歌名" value="{{song.
title}}" />
6         </view>
7         <view>
8           <text>歌手：</text>
```

```
9        <input name="singer" placeholder="请输入歌手" value="{{song.
   singer}}" />
10       </view>
11       <view>
12         <text>歌曲 id（来自网易云音乐网址的 id）：</text>
13         <input name="src" placeholder="请输入歌曲 id" value="{{song.
   src}}" />
14       </view>
15       <view>
16         <text>图片：</text>
17         <image src="{{img}}" mode="aspectFit" bindtap="changeImg"></image>
18       </view>
19       <view>
20         <text>栏目：</text>
21         <checkbox-group name="category">
22           <label wx:for="{{categories}}" wx:key="id">
23             <checkbox value="{{item.id}}" checked="{{indexOf(song.
   category,item.id)>=0}}" /> {{item.name}}
24           </label>
25         </checkbox-group>
26       </view>
27       <button form-type="submit">保存</button>
28     </form>
29   </view>
30
31   <view class="page-foot">本项目所用音乐及歌词来自网易云音乐</view>
32   <wxs module="indexOf">
33     module.exports = function (arr, item) {
34       return arr.indexOf(item)
35     }
36   </wxs>
```

第 31 行中的样式.page-foot 是全局样式，需要添加在 app.wxss 文件中。打开 app.wxss 文件，插入以下代码：

```
1   .page-foot {
2     margin: 100rpx 0 30rpx 0;
3     text-align: center;
4     color: #ccc;
5     font-size: xx-small;
6   }
```

打开 song.wxss 文件，输入以下代码：

```
1   page {
2     background: #21445a;
3     color: #e9eaec;
4     height: 100%;
```

```
5      }
6
7      .container {
8        padding: 50rpx;
9      }
10
11     view {
12       margin-bottom: 30rpx;
13     }
14
15     input {
16       width: 650rpx;
17       margin-top: 10rpx;
18       border-bottom: 2rpx solid #e9eaec;
19     }
20
21     label {
22       display: block;
23       margin: 8rpx;
24     }
25
26     image {
27       width: 650rpx;
28       border-radius: 20rpx;
29     }
```

打开 song.js 文件，删除原来的代码，输入以下代码：

```
1      const util = require('../../utils/util.js')
2
3      let that
4      Page({
5
6        data: {
7          categories: [],
8          song: {},
9          img: '/images/no_image.png'
10       },
11
12       getSongById: function (id) {
13         wx.request({
14           url: util.server_url + 'getSong',
15           method: 'post',
16           data: { id: id },
17           success: res => {
18             console.log(res.data)
19             if (res.data.err_code == 0) {
```

```
20              that.img = res.data.song.img
21              that.setData({
22                song: res.data.song,
23                img: util.file_url + that.img
24              })
25            } else if (res.data.err_code == 3) {
26              util.showMessage('歌曲不存在')
27            }
28          },
29          fail: err => {
30            console.error(err)
31          }
32        })
33      },
34
35      getAllCategories: function (id) {
36        wx.request({
37          url: util.server_url + 'getAllCategories',
38          success: res => {
39            console.log(res.data)
40            if (res.data.err_code == 0) {
41              res.data.categories.forEach(c => {
42                c.imgUrl = util.file_url + c.img
43              });
44              that.setData({
45                categories: res.data.categories
46              })
47            }
48          },
49          fail: err => {
50            console.error(err)
51          }
52        })
53      },
54
55      id: '',
56      img: '',
57      eventChannel: null,
58      onLoad(options) {
59        that = this
60        this.eventChannel = this.getOpenerEventChannel()
61        if (options.id) {
62          this.id = options.id
63        }
64        if (options.id) {
65          this.getSongById(options.id)
```

```
66        }
67        this.getAllCategories()
68      },
69
70      changeImg: function (e) {
71        util.uploadMedia('image',
72          res => {
73            console.log(res)
74            that.img = res.file_path
75            that.setData({ img: util.file_url + res.file_path })
76          },
77          err => {
78            console.error(err)
79          })
80      },
81
82      submit: function (e) {
83        console.log(e)
84        var data = e.detail.value
85        data.img = this.img
86        //song.category 是使用逗号分隔的栏目 id，一首歌曲可以属于多个栏目
87        //原来的 data.category 是一个数组
88        data.category = data.category.join(',')
89        if (this.id) {
90          data.id = this.id
91          wx.request({
92            url: util.server_url + 'updateSong',
93            method: 'post',
94            data: { data: data },
95            success: res => {
96              console.log(res.data)
97              if (res.data.err_code == 0) {
98                console.log(that.eventChannel)
99                that.eventChannel.emit('refresh', { data: true })
100               wx.navigateBack()
101             } else if (res.data.err_code == 2) {
102               util.showMessage('更新歌曲失败')
103             }
104           },
105           fail: err => {
106             console.error(err)
107           }
108         })
109       } else {
110         wx.request({
111           url: util.server_url + 'addSong',
```

```
112            method: 'post',
113            data: { data: data },
114            success: res => {
115              console.log(res.data)
116              if (res.data.err_code == 0) {
117                console.log(res)
118                that.eventChannel.emit('refresh', { data: true })
119                wx.navigateBack()
120              } else if (res.data.err_code == 1) {
121                util.showMessage('增加歌曲失败')
122              }
123            },
124            fail: err => {
125              console.error(err)
126            }
127          })
128        }
129      },
130
131    })
```

song.js 文件的代码与 category.js 文件类似，不再赘述。

至此，完成了编辑栏目和音乐的代码。现在可以运行程序，并添加一些栏目和音乐，方便后续实现音乐播放主界面及音乐播放器界面。

在添加歌曲时，注意歌曲 id 来自网易云音乐网址的 id。以《罗刹海市》（刀郎）为例，这首歌曲的网易云音乐 id 是 2063487880。

任务八　实现音乐播放主界面

扫一扫

微课：媒体播放-首页代码

index 页面实现音乐播放主界面，该页面显示所有的栏目和歌曲。

打开 index.wxml 文件，删除原来的代码，输入以下代码：

```
1     <view class="box" wx:for="{{categories}}" wx:for-index="cIndex" wx:for-item="category" wx:key="id">
2       <view class="category">
3         <image src="{{category.imgUrl}}" mode="aspectFit" />
4         <view class="list-category">{{category.name}}</view>
5       </view>
6       <view class="song" wx:for="{{category.songs}}" wx:for-index="sIndex" wx:for-item="song"
7         wx:key="id" data-id="{{sIndex}}" id="{{cIndex}}" bindtap="play">
8         <image src="{{song.imgUrl}}" mode="aspectFit" />
9         <view class="list-song">
```

```
10            <view>{{song.title}}</view>
11            <view>{{song.singer}}</view>
12          </view>
13        </view>
14      </view>
15
16      <view class="page-foot">本项目所用音乐及歌词来自网易云音乐</view>
```

第 1 行中的列表渲染和第 7 行中的列表渲染形成嵌套，类似编程语言中的双重循环。第 1 行中的 wx:for-item 指定外部列表渲染的项目名称为 category，默认为 item，为了避免和内部的列表渲染重名，需要更换名称。第 7 行中的列表渲染使用的数组是 category.songs，即栏目下的所有歌曲。通过嵌套的列表渲染，使用 16 行代码就显示了所有的栏目和歌曲。

打开 index.wxss 文件，输入以下代码：

```
1     page {
2       background: #21445a;
3       color: #e9eaec;
4       height: 100%;
5     }
6
7     .box {
8       width: 718rpx;
9       margin: 30rpx auto;
10      border-radius: 18rpx;
11      box-shadow: 10rpx 10rpx 10rpx 5rpx rgba(0, 0, 0, 0.4);
12      white-space: nowrap;
13      background-color: #e9eaec;
14      color: #21445a;
15    }
16
17    .category {
18      margin-bottom: 10rpx;
19      display: flex;
20      align-items: center;
21      padding: 20rpx;
22    }
23
24    .list-category {
25      flex: 1;
26      margin-left: 10rpx;
27    }
28
29    .category>image {
30      width: 100rpx;
31      height: 100rpx;
32      border-radius: 20rpx;
```

```
33    }
34
35    .song {
36      margin-bottom: 10rpx;
37      display: flex;
38      align-items: center;
39      padding: 20rpx 20rpx 20rpx 100rpx;
40      font-size: small;
41      border-top: 2rpx dashed #255a5f;
42    }
43
44    .list-song {
45      flex: 1;
46      margin-left: 10rpx;
47    }
48
49    .song>image {
50      width: 100rpx;
51      height: 100rpx;
52      border-radius: 20rpx;
53    }
```

打开 index.json 文件，添加代码"enablePullDownRefresh": true，激活下拉刷新功能。

打开 index.js 文件，删除原来的代码，输入以下代码：

```
1     const util = require('../../utils/util.js')
2
3     let that
4     Page({
5       data: {
6         categories: []
7       },
8       //查找所有栏目，根据栏目 id 查找歌曲，并存放到栏目中一起在界面中显示
9       getAllCategories: function (id) {
10        wx.request({
11          url: util.server_url + 'getAllCategories',
12          success: res => {
13            console.log(res.data)
14            if (res.data.err_code == 0) {
15              let categories = res.data.categories
16              categories.forEach(c => {
17                c.imgUrl = util.file_url + c.img
18                //根据栏目 id 查找歌曲
19                wx.request({
20                  url: util.server_url + 'getSongsByCategory?id=' + c.id,
21                  success: res => {
22                    console.log(res.data)
```

```
23                  if (res.data.err_code == 0) {
24                    res.data.songs.forEach(s => {
25                      s.imgUrl = util.file_url + s.img
26                    })
27                    c.songs = res.data.songs
28                    //不可以在 forEach 函数后面调用 setData 函数
29                    that.setData({
30                      categories: categories
31                    })
32                  }
33                },
34                fail: err => {
35                  console.error(err)
36                }
37              })
38            })
39          }
40        },
41        fail: err => {
42          console.error(err)
43        }
44      })
45    },
46
47    onLoad: function (options) {
48      that = this
49      this.getAllCategories()
50    },
51
52    onPullDownRefresh() {
53      this.getAllCategories()
54    },
55
56    play: function (e) {
57      const sIndex = e.currentTarget.dataset.id
58      const cIndex = e.currentTarget.id
59      wx.navigateTo({
60        url: '/pages/play/play',
61        success: function (res) {
62          // 通过 eventChannel 向打开的页面传递数据
63          res.eventChannel.emit('songs', {
64            songs: that.data.categories[cIndex].songs,
65            sIndex: sIndex
66          })
67        }
68      })
```

```
69        }
70      })
```

第 56 行中的 play 方法是单击某首歌曲触发的事件处理方法，第 59 行表示跳转到下一个任务完成的音乐播放器页面 play，第 63～66 行通过 eventChannel 发送指定栏目下的所有歌曲和单击歌曲的下标 Index 到 play 页面去播放。

任务九　实现音乐播放器界面

play 页面实现音乐播放器，该页面中间是音乐播放器，向右滑动屏幕显示歌词，向左滑动屏幕显示播放的歌曲列表。页面底部是播放工具条，显示五个按钮，分别是“播放模式”按钮（顺序播放、随机播放、单曲循环）、“上一首”按钮、“播放”/“暂停”按钮、“下一首”按钮、“播放列表”按钮。

扫一扫

微课：媒体播放-播放器代码 1

打开 play.wxml 文件，删除原来的代码，输入以下代码：

```
1     <swiper current="{{current}}">
2       <swiper-item>
3         <!-- 歌词 -->
4         <scroll-view class="lyric" scroll-y scroll-into-view=
      "lyric{{lyricIndex-10>=0?lyricIndex-10:0}}" scroll-with-animation>
5           <text id="lyric{{index}}" class="content-play-lyric{{index==
      lyricIndex?'-active':''}}" wx:for="{{lyric}}" wx:key="time">{{item.
      text}}</text>
6         </scroll-view>
7       </swiper-item>
8       <swiper-item>
9         <!-- 播放器 -->
10        <view class="play">
11          <!-- 歌曲名、歌手 -->
12          <view class="play-info">
13            <view>{{song.title}}</view>
14            <view>{{song.singer}}</view>
15          </view>
16          <view class="cd">
17            <image src="/images/cd.png" />
18          </view>
19          <!-- 唱片 -->
20          <view class="play-img play-img-{{playing ? 'play' : 'pause'}}">
21            <image src="{{song.imgUrl}}" />
22            <image src="/images/pin7.png" />
23          </view>
24          <!-- 播放进度 -->
```

```
25            <view class="progress">
26              <text style="padding: 0 20rpx;">{{progress.currentTime}}</text>
27              <view>
28                <slider bindchange="sliderChange" activeColor="#d33a31"
   block-size="12" backgroundColor="#dadada" value="{{progress.percent}}" />
29              </view>
30              <text style="padding: 0 20rpx;">{{progress.duration}}</text>
31            </view>
32          </view>
33        </swiper-item>
34        <swiper-item>
35          <!-- 歌曲列表 -->
36          <view class="list-empty" wx:if="{{songs.length === 0}}">没有歌曲
   </view>
37          <scroll-view scroll-y="true" class="scrolly">
38            <view class="list-item" wx:for="{{songs}}" wx:key="id"
   bindtap="playSong" data-id="{{index}}">
39              <image class="list-img" src="{{item.imgUrl}}" />
40              <view class="list-info {{songIndex == index ? 'active': ''}}">
41                <view>{{item.title}}</view>
42                <view>{{item.singer}}</view>
43              </view>
44            </view>
45          </scroll-view>
46        </swiper-item>
47      </swiper>
48      <!-- "播放"按钮 -->
49      <view class="controls">
50        <image src="/images/{{playMode==0?'sequence':playMode==1?'random':
   'cycle'}}.png" bindtap="playMode" />
51        <image src="/images/prev.png" bindtap="prev" />
52        <image wx:if="{{!playing}}" src="/images/play.png" bindtap="play" />
53        <image wx:else src="/images/pause.png" bindtap="pause" />
54        <image src="/images/next.png" bindtap="next" />
55        <image src="/images/list.png" bindtap="songList" />
56      </view>
```

页面由一个滑块视图容器 swiper 和底部的“播放”按钮组成。第 50 行表示通过 playMode 显示对应的图片，如果 playMode 是 0，则代表顺序播放，显示 sequence.png；如果 playMode 是 1，则代表随机播放，显示 random.png；如果 playMode 是 2，则代表单曲循环，显示 cycle.png。

扫一扫

微课：媒体播放-播放器代码 2

打开 play.wxss 文件，输入以下代码：

```
1    page {
2      background: #21445a;
```

```
3      color: #e9eaec;
4      height: 100%;
5    }
6
7    swiper {
8      height: 90vh;
9    }
10
11   .play {
12     position: relative;
13     height: 100%;
14   }
15
16   /* 歌名、歌手 */
17   .play-info {
18     text-align: center;
19     margin: 50rpx;
20   }
21
22   .play-info>view:last-child {
23     font-size: smaller;
24   }
25
26   /* 唱片图片和唱针图片 */
27   .play-img {
28     position: relative;
29   }
30
31   .cd {
32     width: 450rpx;
33     height: 450rpx;
34     position: absolute;
35     top: 270rpx;
36     left: 120rpx;
37     border-radius: 50%;
38   }
39
40   .cd>image {
41     width: 100%;
42     height: 100%;
43     border-radius: 50%;
44     border: 30rpx solid #555;
45   }
46
47   .play-img>image:first-child {
48     width: 350rpx;
```

```
49      height: 350rpx;
50      position: absolute;
51      top: 215rpx;
52      left: 200rpx;
53      border-radius: 50%;
54      animation: cdRotate 10s linear infinite;
55    }
56
57    @keyframes cdRotate {
58      from {
59        transform: rotate(0deg);
60      }
61
62      to {
63        transform: rotate(360deg);
64      }
65    }
66
67    .play-img>image:last-child {
68      width: 147rpx;
69      height: 348rpx;
70      position: absolute;
71      top: 0;
72      left: 270rpx;
73    }
74
75    .play-img-play>image:first-child {
76      animation-play-state: running;
77    }
78
79    .play-img-play>image:last-child {
80      animation: musicStart 0.2s linear forwards;
81      transform-origin: 110rpx 30rpx;
82    }
83
84    .play-img-pause>image:first-child {
85      animation-play-state: paused;
86    }
87
88    .play-img-pause>image:last-child {
89      animation: musicStop 0.2s linear forwards;
90      transform-origin: 110rpx 30rpx;
91    }
92
93    @keyframes musicStart {
94      from {
```

```
95        transform: rotate(50deg);
96      }
97
98      to {
99        transform: rotate(23deg);
100     }
101   }
102
103   @keyframes musicStop {
104     from {
105       transform: rotate(23deg);
106     }
107
108     to {
109       transform: rotate(50deg);
110     }
111   }
112
113   /* 进度条 */
114   .progress {
115     width: 100%;
116     position: absolute;
117     bottom: 50rpx;
118     left: 0rpx;
119     display: flex;
120     align-items: center;
121     font-size: 9pt;
122     text-align: center;
123   }
124
125   .progress>view {
126     flex: 1;
127   }
128
129   /*"播放"按钮 */
130   .controls {
131     display: flex;
132     align-items: center;
133     height: 120rpx;
134     justify-content: space-around;
135   }
136
137   .controls>image {
138     width: 80rpx;
139     height: 80rpx;
140   }
```

```
141
142     /* 歌词 */
143     .lyric {
144       height: 95%;
145       padding: 20rpx;
146       text-align: center;
147     }
148
149     .content-play-lyric {
150       color: #e9eaec;
151       font-size: 11pt;
152       margin: 5rpx;
153       display: block;
154     }
155
156     .content-play-lyric-active {
157       color: #ce8181;
158       font-size: 12pt;
159       margin: 5rpx;
160       display: block;
161     }
162
163     /* 歌曲列表 */
164     .list-item {
165       display: flex;
166       align-items: center;
167       border: 1rpx solid #e9eaec;
168       margin: 20rpx;
169       border-radius: 10rpx;
170       box-shadow: 10rpx 10rpx 10rpx 5rpx rgba(0, 0, 0, 0.4);
171       height: 130rpx;
172     }
173
174     .list-empty {
175       margin: 80rpx, auto;
176       padding: 50rpx;
177       text-align: center;
178     }
179
180     .list-img {
181       width: 80rpx;
182       height: 80rpx;
183       margin-left: 15rpx;
184       border-radius: 8rpx;
185       border: 1px solid #21445a;
186     }
```

```
187
188    .list-info {
189      flex: 1;
190      font-size: 10pt;
191      line-height: 38rpx;
192      margin-left: 20rpx;
193      padding-bottom: 8rpx;
194    }
195
196    .list-info.active {
197      color: #ce8181;
198    }
```

第 47～111 行实现了两种动画显示：唱片的无限循环旋转和唱针的指定角度旋转。唱片对应“.play-img>image:first-child”，唱针对应“.play-img>image:last-child”，它们的动画实现方式不同。

（1）唱片旋转。第 54 行中的“cdRotate 10s linear infinite”指定动画的形式分别为 cdRotate（在第 57～65 行定义）、10 秒完成一次、匀速、无限循环。第 57～65 行定义 cdRotate 采用关键帧，从 0 到 360 度，因此顺时针旋转一周，结合无限循环就可以实现唱片无限顺时针循环旋转。第 75～77 行和第 84～86 行分别用来控制播放动画和暂停动画，结合 js 代码实现在播放歌曲时旋转动画，在暂停播放歌曲时暂停动画。

（2）唱针旋转。第 79～82 行定义了播放歌曲时唱针的动画，第 80 行中的“musicStart 0.2s linear forwards”指定动画的形式分别为 musicStart、0.2 秒完成一次、匀速、向前，代表只执行一次动画。第 93～101 行定义了 musicStart 关键帧，从 50 度旋转到 23 度（逆时针旋转）。第 81 行定义了旋转的中心坐标。类似地，第 88～91 行定义了暂停播放歌曲时唱针的动画，第 103～111 行定义了 musicStop 关键帧，不再赘述。结合 js 代码实现在播放歌曲时唱针逆时针旋转，“别”在唱片上，在暂停播放歌曲时，唱针顺时针旋转，离开唱片。

扫一扫

微课：媒体播放-播放器代码 3

打开 play.js 文件，删除原来的代码，输入以下代码：

```
1    const util = require('../../utils/util.js')
2    const path = 'https://music.163.com/song/media/outer/url?id=' //网易云音乐外链链接
3    const lyricPath = 'https://music.163.com/api/song/media?id=' //歌词链接
4    let that
5
6    Page({
7      data: {
8        current: 1,//swiper.current
9        song: {
10         imgUrl: '/images/cd.png',
11         title: '流浪地球',
```

```
12        singer: '吴京'
13      },
14      lyric: [],
15      lyricIndex: -1,
16      progress: {
17        currentTime: '00:00',
18        percent: 50,
19        duration: '00:00'
20      },
21      songs: [],
22      songIndex: 0,
23      playing: false,
24      playMode: 0//播放模式：0 代表顺序播放，1 代表随机播放，2 代表单曲循环
25    },
26
27    onLoad(options) {
28      that = this
29      //接收 index 页面发送的数据
30      const eventChannel = this.getOpenerEventChannel()
31      eventChannel.on('songs', function (data) {
32        console.log(data)
33        that.setData({
34          songs: data.songs,
35          songIndex: data.sIndex,
36        })
37      })
38    },
39
40    audio: null,
41    onReady: function () {
42      //使用背景音乐播放音乐，需要在 app.json 文件中配置
43      //"requiredBackgroundModes": ["audio"],
44      this.audio = wx.getBackgroundAudioManager()
45      this.audio.onError(function () {
46        console.log('播放失败：' + that.audio.src)
47      })
48      //播放完成自动换下一曲
49      this.audio.onEnded(function () {
50        that.next()
51      })
52      this.audio.onTimeUpdate(this.updateTime)
53      //修改背景音乐默认的自动播放
54      this.audio.onCanplay(() => {
55        this.audio.pause()
56      })
57      this.setSong()
```

```
58      },
59
60      //格式化时间
61      formatTime: function (time) {
62        var minute = Math.floor(time / 60) % 60;
63        var second = Math.floor(time) % 60
64        return (minute < 10 ? '0' + minute : minute) + ':' + (second <
        10 ? '0' + second : second)
65      },
66      //在模拟器中的显示不太正常，真机预览没有问题
67      updateTime: function () {
68        this.setLyricIndex(this.audio.currentTime)
69        this.setData({
70          'progress.duration': this.formatTime(this.audio.duration),
71          'progress.currentTime':
        this.formatTime(this.audio.currentTime),
72          'progress.percent': this.audio.currentTime / this.audio.
        duration * 100
73        })
74      },
75
76      setSong: function () {
77        const song = this.data.songs[this.data.songIndex]
78        //获取歌曲图片服务器的完整链接
79        song.imgUrl = util.file_url + song.img
80        //网易云音乐外链链接
81        this.audio.src = path + song.src + '.mp3'
82        this.audio.title = song.title
83        this.audio.epname = 'music'
84        this.audio.singer = song.singer
85        this.audio.coverImgUrl = util.file_url + song.img
86        //获取歌词
87        wx.request({
88          url: lyricPath + song.src,
89          success: res => {
90            console.log(res.data)
91            if (res.data.lyric) {
92              this.formatLyric(res.data.lyric)
93              console.log(that.data.lyric)
94            }
95          }
96        })
97        this.setData({
98          'progress.currentTime': '00:00',
99          'progress.duration': '00:00',
100         'progress.percent': 0,
```

```
101           lyricIndex: -1,
102           song: song
103         })
104       },
105
106       //传入初始歌词文本text
107       formatLyric(text) {
108         let result = []
109         //原歌词文本已经换行，直接通过换行符“\n”进行切割
110         let arr = text.split("\n")
111         let row = arr.length//获取歌词行数
112         for (let i = 0; i < row; i++) {
113           let temp_row = arr[i]//每行格式“ [00:04.302][02:10.00]hello
    worLd”
114           let temp_arr = temp_row.split("]")//通过“]”对时间和文本进行分离
115           //把歌词文本从数组中剔除，获取歌词文本
116           //对剩下的歌词时间进行处理
117           let text = temp_arr.pop()
118           temp_arr.forEach(element => {
119             let obj = {}
120             //先把多余的“[”去掉，再分离出分、秒
121             let time_arr = element.substr(1, element.length - 1).split(":")
122             //把时间转换成与currentTime相同的类型，方便后续实现
123             let s = parseInt(time_arr[0]) * 60 + parseFloat(time_arr[1])
124             obj.time = s
125             obj.text = text
126             if (text != "")
127               result.push(obj)//将每行歌词对象存放到组件的lyric歌词属性中
128           })
129         }
130         //因为把不同时间的相同歌词排列在一起，所以以时间顺序重新排列
131         result.sort(this.sortRule)
132         this.setData({ lyric: result })
133       },
134
135       sortRule(a, b) {//设置排列规则
136         return a.time - b.time
137       },
138
139       //根据播放进度设置当前歌词的index页面
140       setLyricIndex(time) {
141         for (var i = 0; i < this.data.lyric.length; i++) {
142           //如果某行歌词的播放时间大于time，则显示上一行歌词
143           if (this.data.lyric[i].time > time) {
144             //避免同一行歌词反复
145             if (this.data.lyricIndex != i - 1)
```

```
146             this.setData({ lyricIndex: i - 1 })
147           return
148         } else { //已经到最后一段歌词
149           if (i == this.data.lyric.length - 1) {
150             //避免同一行歌词反复
151             if (this.data.lyricIndex != this.data.lyric.length - 1)
152               this.setData({ lyricIndex: this.data.lyric.length - 1 })
153             return
154           }
155         }
156       }
157     },
158 
159     play: function () {
160       //恢复自动播放
161       this.audio.onCanplay(() => { })
162       this.audio.play()
163       this.setData({ playing: true })
164     },
165 
166     pause: function () {
167       this.audio.pause()
168       this.setData({ playing: false })
169     },
170 
171     next: function () {
172       let songIndex = this.data.songIndex
173       if (this.data.playMode == 0) {//顺序播放
174         songIndex = songIndex >= this.data.songs.length - 1 ? 0 :
    songIndex + 1
175       } else if (this.data.playMode == 1) {//随机播放
176         songIndex = Math.floor(Math.random() * this.data.songs.length)
177         console.log(songIndex)
178       } //单曲循环，songIndex 不变
179       this.setData({ songIndex: songIndex })
180       this.setSong()
181       if (this.data.playing) {
182         this.play()
183       }
184     },
185 
186     prev: function () {
187       let songIndex = this.data.songIndex
188       if (this.data.playMode == 0) {//顺序播放
189         songIndex = songIndex == 0 ? this.data.songs.length - 1 :
    songIndex - 1
```

```
190        } else if (this.data.playMode == 1) {//随机播放
191          songIndex = Math.floor(Math.random() * this.data.songs.length)
192        } //单曲循环，songIndex 不变
193        this.setData({ songIndex: songIndex })
194        this.setSong()
195        if (this.data.playing) {
196          this.play()
197        }
198      },
199
200      playMode: function (e) {
201        let playMode = this.data.playMode
202        playMode++
203        if (playMode == 3) playMode = 0
204        this.setData({ playMode: playMode })
205      },
206
207      songList: function (e) {
208        let current = this.data.current
209        current = current == 1 ? 2 : 1
210        this.setData({ current: current })
211      },
212
213      sliderChange: function (e) {
214        var second = e.detail.value * this.audio.duration / 100
215        this.audio.seek(second)
216        //解决单击进度条不更新数字的问题
217        this.updateTime()
218      },
219
220      playSong: function (e) {
221        this.setData({ songIndex: e.currentTarget.dataset.id })
222        this.setSong()
223        this.play()
224      },
225    })
```

第 31～37 行接收 index 页面发送的消息，消息携带歌曲列表和当前歌曲的下标。第 52 行设置监听背景音乐播放进度更新事件为 updateTime 方法，只要小程序在前台播放音乐，就一直调用该方法。第 67～74 行中的 updateTime 方法更新当前播放的歌词及进度条。第 76～104 行中的 setSong 方法用来设置当前播放歌曲，第 87～96 行用来向网易云音乐获取歌词。

第 107～137 行用于处理网易云音乐返回的歌词，分离出播放时间段和歌词，获取包括时间和歌词的列表，并保存在 data.lyric 中，方便后面显示当前时间应该播放的歌词，具体解释见代码中的注释。

第 140～157 行中的 setLyricIndex 方法在 updateTime 方法中调用，用于设置当前正在播放的歌词。根据当前播放时间，遍历包括时间和歌词的列表，获取当前播放的歌词下标，并保存在 data.lyricIndex 中，具体解释见代码中的注释。

第 191 行中的 Math.floor(Math.random() * this.data.songs.length)调用了两个 Math 的方法，floor 表示返回小于或等于给定数字的最大整数，即向下取整；Math.random()返回 0（包含）～1（不包含）之间的随机数，Math.random() * this.data.songs.length 返回 0（包含）～length（不包含）之间的浮点数。因此 Math.floor(Math.random() * this.data.songs.length)表示返回 0～length-1 之间的随机整数，即对应歌曲的下标。

至此音乐播放器已经实现。

任务十　编辑视频

扫一扫

微课：媒体播放-视频代码

打开 admin.wxml 文件，将光标移动到“视频”一行，按 Ctrl+/注释这行，并在后面输入以下代码：

```
1         <swiper-item>
2           <!-- 视频 -->
3           <view class="list-empty" wx:if="{{videos.length === 0}}">没有视
  频</view>
4           <scroll-view scroll-y="true" class="scrolly">
5             <view class="list-item" wx:for="{{videos}}" wx:key="id"
  bindtap="video" data-id="{{item.id}}">
6               <view class="list-info">{{item.name}}</view>
7               <image class="arrow" src="/images/arrow.png"></image>
8             </view>
9           </scroll-view>
10        </swiper-item>
```

打开 admin.js 文件，在 data{}的后面插入以下代码：

```
    videos: []
```

在 onLoad 方法的后面插入以下代码：

```
    this.getAllVideos()
```

在 add 方法的 case 2 代码段后面插入以下代码：

```
1           case 2:
2             wx.navigateTo({
3               url: '/pages/video/video',
4               events: {
5                 refresh: function (data) {
6                   that.getAllVideos()
```

```
7              }
8            }
9          })
10         break
```

在 Page({})的后面插入 getAllVideos 和 video 两个方法，代码如下：

```
1      getAllVideos: function () {
2        wx.request({
3          url: util.server_url + 'getAllVideos',
4          success: res => {
5            console.log(res.data)
6            if (res.data.err_code == 0) {
7              that.setData({
8                videos: res.data.videos
9              })
10           }
11         },
12         fail: err => {
13           console.error(err)
14         }
15       })
16     },
17
18     video: function (e) {
19       console.log(e)
20       wx.navigateTo({
21         url: '/pages/video/video?id=' +
22           e.currentTarget.dataset.id,
23         events: {
24           refresh: function (data) {
25             that.getAllVideos()
26           }
27         },
28       })
29     },
```

getAllVideos 方法从服务器读取所有的视频数据。video 方法是单击某个视频后，跳转到 video 页面进行修改。增加视频和修改视频都要跳转到 video 页面，使用方法与栏目的编辑相同，不再赘述。

接下来编写 video 页面的代码。打开 video.wxml 文件，删除原来的代码，输入以下代码：

```
1    <view class="container">
2      <form bindsubmit="submit">
3        <view>
4          <text>视频名称：</text>
```

```
5            <input name="name" placeholder="请输入视频名称" value="{{video.
   name}}" />
6          </view>
7          <view>
8            <text>视频来源：</text>
9            <input name="source" placeholder="请输入视频来源" value="{{video.
   source}}" />
10         </view>
11         <view>
12           <text>视频：</text>
13           <video src="{{src}}" controls></video>
14           <button bindtap="changeVideo">上传视频</button>
15         </view>
16         <button form-type="submit">保存</button>
17       </form>
18     </view>
```

打开 video.wxss 文件，输入以下代码：

```
1     page {
2       background: #21445a;
3       color: #e9eaec;
4       height: 100%;
5     }
6
7     .container {
8       padding: 50rpx;
9     }
10
11    view {
12      margin-bottom: 30rpx;
13    }
14
15    input {
16      width: 650rpx;
17      margin-top: 10rpx;
18      border-bottom: 2rpx solid #e9eaec;
19    }
20
21    image {
22      width: 650rpx;
23      border-radius: 20rpx;
24    }
```

打开 video.js 文件，删除原来的代码，输入以下代码：

```
1     const util = require('../../utils/util.js')
2
```

```
3    let that
4    Page({
5
6      data: {
7        video: {},
8        src: ''
9      },
10
11     getVideoById: function (id) {
12       wx.request({
13         url: util.server_url + 'getVideo',
14         method: 'post',
15         data: { id: id },
16         success: res => {
17           console.log(res.data)
18           if (res.data.err_code == 0) {
19             that.src = res.data.video.src
20             that.setData({
21               video: res.data.video,
22               src: util.file_url + that.src
23             })
24           } else if (res.data.err_code == 3) {
25             util.showMessage('视频不存在')
26           }
27         },
28         fail: err => {
29           console.error(err)
30         }
31       })
32     },
33
34     id: '',
35     src: '',
36     eventChannel: null,
37     onLoad(options) {
38       that = this
39       this.eventChannel = this.getOpenerEventChannel()
40       if (options.id) {
41         this.id = options.id
42       }
43       if (options.id) {
44         this.getVideoById(options.id)
45       }
46     },
47
48     changeVideo: function (e) {
```

```
49         util.uploadMedia('video',
50           res => {
51             console.log(res)
52             that.src = res.file_path
53             that.setData({ src: util.file_url + res.file_path })
54           },
55           err => {
56             console.error(err)
57           })
58       },
59 
60       submit: function (e) {
61         console.log(e)
62         var data = e.detail.value
63         data.src = this.src
64         if (this.id) {
65           data.id = this.id
66           wx.request({
67             url: util.server_url + 'updateVideo',
68             method: 'post',
69             data: { data: data },
70             success: res => {
71               console.log(res.data)
72               if (res.data.err_code == 0) {
73                 console.log(that.eventChannel)
74                 that.eventChannel.emit('refresh', { data: true })
75                 wx.navigateBack()
76               } else if (res.data.err_code == 2) {
77                 util.showMessage('更新视频失败')
78               }
79             },
80             fail: err => {
81               console.error(err)
82             }
83           })
84         } else {
85           wx.request({
86             url: util.server_url + 'addVideo',
87             method: 'post',
88             data: { data: data },
89             success: res => {
90               console.log(res.data)
91               if (res.data.err_code == 0) {
92                 console.log(res)
93                 that.eventChannel.emit('refresh', { data: true })
94                 wx.navigateBack()
```

```
95              } else if (res.data.err_code == 1) {
96                util.showMessage('增加视频失败')
97              }
98            },
99            fail: err => {
100             console.error(err)
101           }
102         })
103       }
104     },
105 
106   })
```

本段代码与 category.js 文件和 song.js 文件中的类似，不再赘述。

至此完成编辑视频页面的代码。

任务十一　播放视频并发送弹幕

播放视频采用向上滑动的方式进行视频切换。因此还是使用滑块视图容器 swiper 来实现，只是改为纵向滑动。

打开 play_video.wxml 文件，删除原来的代码，输入以下代码：

```
1   <swiper current="{{vIndex}}" vertical bindchange="changeVideo">
2    <swiper-item wx:for="{{videos}}" wx:key="id">
      <video id="video{{index}}" bindtimeupdate="timeupdate" src=
3  "{{item.srcUrl}}" danmu-list="{{barrages}}" enable-danmu danmu-btn
   controls></video>
4     <view class="info">视频名称：{{item.name}}</view>
5     <view class="info">视频来源：{{item.source}}</view>
6     <view class="barrage">
7      <input placeholder="请输入弹幕" bindblur="change" />
8      <image src="/images/barrage.png" bindtap="addBarrage" />
9     </view>
10   </swiper-item>
11  </swiper>
```

第 1 行增加了 vertical 属性，滑块视图容器纵向滑动。第 3 行为每个 video 设置了对应的 id，如第 0 个 video 的 id 是 video0。第 7 行的 bindblur 是失去焦点时触发的事件，即当输入焦点离开 input，如单击发送图片时，就触发。

打开 play_video.wxss 文件，输入以下代码：

```
1   page {
2     background: #21445a;
3     color: #e9eaec;
```

```
4      height: 100%;
5    }
6
7    swiper {
8      height: 100%;
9    }
10
11   video {
12     width: 100%;
13   }
14
15   .info {
16     text-align: center;
17     margin: 0 20rpx;
18   }
19
20   .barrage {
21     display: flex;
22     align-items: center;
23     height: 100rpx;
24     padding: 0 20rpx;
25   }
26
27   .barrage>input {
28     flex: 1;
29     margin-top: 10rpx;
30     border-bottom: 2rpx solid #e9eaec;
31   }
32
33   .barrage>image {
34     width: 80rpx;
35     height: 80rpx;
36   }
```

打开 play_video.json 文件，添加代码"enablePullDownRefresh": true，激活下拉刷新功能。

打开 play_video.js 文件，删除原来的代码，输入以下代码：

```
1    const util = require('../../utils/util.js')
2
3    let that
4
5    function getRandomColor () {
6      let rgb = []
7      for (let i = 0 ; i < 3; ++i) {
8        //因为弹幕是黑底，所以颜色要浅，设置所有 rgb>128
9        let color = Math.floor(Math.random() * 128 + 128).toString(16)
10       color = color.length == 1 ? '0' + color : color
```

```
11          rgb.push(color)
12        }
13        return '#' + rgb.join('')
14      }
15      Page({
16
17        data: {
18          videos: [],
19          vIndex: 0,
20          barrages: []
21        },
22
23        getAllVideos: function () {
24          wx.request({
25            url: util.server_url + 'getAllVideos',
26            success: res => {
27              console.log(res.data)
28              if (res.data.err_code == 0) {
29                //拼接完整的视频路径
30                res.data.videos.forEach(v => {
31                  v.srcUrl = util.file_url + v.src
32                })
33                that.setData({
34                  videos: res.data.videos,
35                  vIndex: 0
36                })
37                if (res.data.videos.length > 0) {
38                  that.getBarragesByVideoId(res.data.videos[0].id)
39                  that.videoContext = wx.createVideoContext('video0')
40                  //每个视频都显示当前播放进度
41                  that.currentTimes.length = res.data.videos.length
42                }
43              }
44            },
45            fail: err => {
46              console.error(err)
47            }
48          })
49        },
50
51        getBarragesByVideoId: function (id) {
52          wx.request({
53            url: util.server_url + 'getBarragesByVideoId?id=' + id,
54            success: res => {
55              console.log(res.data)
56              if (res.data.err_code == 0) {
```

```
57            res.data.barrages.forEach(b => {
58              b.color = getRandomColor() //设置弹幕颜色
59            })
60            that.setData({
61              barrages: res.data.barrages
62            })
63          }
64        },
65        fail: err => {
66          console.error(err)
67        }
68      })
69    },
70
71    videoContext: null,
72    onLoad(options) {
73      that = this
74      this.getAllVideos()
75    },
76
77    //下拉刷新视频
78    onPullDownRefresh() {
79      this.getAllVideos()
80    },
81
82    changeVideo: function (e) {
83      let vIndex = e.detail.current
84      this.getBarragesByVideoId(this.data.videos[vIndex].id)
85      this.setData({ vIndex: vIndex })
86      //暂停原来的视频，不同时播放多个视频
87      if(this.videoContext)
88        this.videoContext.pause()
89      this.videoContext = wx.createVideoContext('video' + vIndex)
90    },
91
92    barrage: '',
93    change: function (e) {
94      this.barrage = e.detail.value
95    },
96
97    currentTimes: [],
98
99    //因为保存弹幕需要保存当前视频的播放进度，所以需要处理该事件
100   timeupdate: function(e) {
101     let id = e.currentTarget.id
102     //因为id="video{{index}}"，所以使用substring(5)截取视频index页面
```

```
103        let vIndex = id.substring(5)
104        this.currentTimes[vIndex] = e.detail.currentTime
105      },
106
107      addBarrage: function (e) {
108        let data = {}
109        data.vid = this.data.videos[this.data.vIndex].id
110        data.text = this.barrage
111        data.time = this.currentTimes[this.data.vIndex]
112        wx.request({
113          url: util.server_url + 'addBarrage',
114          method: 'post',
115          data: { data: data },
116          success: res => {
117            console.log(res.data)
118            if (res.data.err_code == 0) {
119              console.log(res)
120              //发送弹幕
121              this.videoContext.sendDanmu({
122                text: that.barrage,
123                color: getRandomColor()
124              })
125            } else if (res.data.err_code == 1) {
126              util.showMessage('保存弹幕失败')
127            }
128          },
129          fail: err => {
130            console.error(err)
131          }
132        })
133      }
134    })
```

第 5 行中的 getRandomColor 方法用来获取随机颜色，作为弹幕颜色。第 9 行的两个 Math 方法用来得到 128～255 的随机整数，作为 RGB 颜色。第 11 行将三个随机的颜色数值以十六进制字符串的形式保存到数组 rgb 中，第 13 行得到对应的颜色字符串并返回。

发送弹幕需要知道当前的播放时间，以后该视频播放到指定时间就自动弹出该弹幕。为了得到当前视频的播放时间，为 video 组件绑定了第 100 行的 timeupdate 事件处理方法。第 104 行将视频播放的当前事件保存在对应下标的 currentTimes 中。在第 111 行中取出发送弹幕的时刻保存的 currentTimes 数组对应下标的值，作为发送弹幕的时刻，并和弹幕文字等一起提交到服务器中保存。

至此媒体播放器项目全部完成，建议同时使用微信开发者工具和手机进行测试。

项目小结

本项目完成的是本教材中难度最大的一个案例，综合性强，具备了背景音乐播放、视频播放、滑块组件使用、tabBar、常见的服务器数据库操作、页面动画使用等常用的微信小程序项目功能。其中的很多代码可以被移植到其他微信小程序项目中。通过学习本项目的案例，可以使学生全面掌握编写较复杂的微信小程序项目的方法。

习题

一、判断题

1．BackgroundAudioManager 对象设置了 src 属性后会自动播放。 （ ）

2．微信小程序的 video 组件需要指定 src 属性使 js 代码获取 VideoContext 实例。 （ ）

3．可以使用 swiper 组件来显示轮播图。 （ ）

二、选择题

1．swiper 组件的（ ）属性是当前所在滑块的下标。

A．index

B．value

C．current

D．interval

2．wx.chooseMedia 指定（ ）参数来确定选择音乐或视频。

A．mediaType

B．sourceType

C．sizeType

D．srcType

3．跳转到非 tabBar 页面，不能使用（ ）方法。

A．wx.switchTab

B．wx.reLaunch

C．wx.navigateTo

D．wx.redirectTo

三、填空题

1．tabBar 的属性________是 tab 按钮的列表。

2．使用 wx.navigateTo 方法跳转页面时，通过___________向跳转页面发送消息。

3．设置 swiper 的______________属性可以实现纵向滑动。

项目六　地点搜索与路线规划

任务一　使用 Map 组件（含获取用户当前位置）

6.1.1　Map 组件属性

外卖类小程序和物流快递类小程序等都会使用与地图相关的应用。微信小程序可以借助 Map 组件实现与地图相关的应用。要在微信小程序中添加地图应用，需要先添加 Map 组件，然后在地图上添加 marker 地图标记、polyline 系列坐标点、controls 显示控件等。表 6.1.1 所示是 Map 组件的主要属性。

表 6.1.1　Map 组件的主要属性

属性	类型	默认值	说明
longitude	number		中心经度
latitude	number		中心纬度
scale	number	16	缩放级别，取值范围为 3～20
markers	array		标记点
autoplay	boolean		是否自动播放
polyline	array		路线
circles	array		圆
rotate	number	0	旋转角度，取值范围为 0～360，地图正北和设备 y 轴角度的夹角
skew	number	0	倾斜角度，取值范围为 0～40，关于 z 轴的倾角
enable-3D	boolean	False	展示 3D 楼块（微信开发者工具暂不支持）
show-compass	boolean	False	显示指南针
show-scale	boolean	False	显示比例尺（微信开发者工具暂不支持）
enable-overlooking	boolean	False	开启俯视
enable-zoom	boolean	True	是否支持缩放
enable-scroll	boolean	True	是否支持拖动

续表

属性	类型	默认值	说明
enable-rotate	boolean	False	是否支持旋转
enable-satellite	boolean	False	是否开启卫星图
enable-traffic	boolean	False	是否开启实时路况
setting	object		配置项
bindcallouttap	eventhandle		单击标记点对应的气泡时触发，e.detail = {markerId}
bindmarkertap	eventhandle		单击标记点时触发，e.detail = {markerId}
bindcontroltap	eventhandle		单击控件时触发，e.detail = {controlId}
bindregionchange	eventhandle		视野发生变化时触发
bindtap	eventhandle		单击地图时触发
bindupdated	eventhandle		在地图渲染更新完成时触发

marker 标记点的属性如表 6.1.2 所示。

表 6.1.2　marker 标记点的属性表

属性	说明	类型	是否必填	备注
id	标记点 id	number	否	marker 单击事件回调会返回此 id。建议为每个 marker 设置 number 类型的 id，以保证更新 marker 时有更好的性能
latitude	纬度	number	是	浮点数，取值范围为-90～90
longitude	经度	number	是	浮点数，取值范围为-180～180
title	标记点名	string	否	单击时显示。callout 存在时它将被忽略
zIndex	显示层级	number		
iconPath	显示的图标	string	是	项目目录下的图片路径。支持相对路径写法，以“/”开头则表示相对小程序根目录，也支持临时路径和网络图片
rotate	旋转角度	number	否	顺时针旋转的角度，取值范围为 0～360，默认为 0
alpha	标注的透明度	number	否	默认为 1，不透明
width	标注图标的宽度	number	否	默认为图片的实际宽度
height	标注图标的高度	number	否	默认为图片的实际高度

如果在地图上进行路线规划，则需要使用坐标点，表 6.1.3 所示是 polyline 坐标点的属性。

表 6.1.3　polyline 坐标点的属性

属性	说明	类型	是否必填	备注
points	经纬度数组	array	是	[{latitude: 0, longitude: 0}]
color	线的颜色	string	否	使用八位十六进制数表示，后两位表示 Alpha 值，如#000000AA

续表

属性	说明	类型	是否必填	备注
width	线的宽度	number	否	
dottedLine	是否为虚线	boolean	否	默认为 False
arrowLine	是否为带箭头的线	boolean	否	默认为 False。微信开发者工具暂不支持该属性
arrowIconPath	更换箭头图标	string	否	在 arrowLine 为 True 时生效
borderColor	线的边框颜色	string	否	
borderWidth	线的厚度	number	否	
latitude	纬度	number	是	浮点数，取值范围为-90～90
longitude	经度	number	是	浮点数，取值范围为-180～180
color	描边的颜色	string	否	使用八位十六进制数表示，后两位表示 Alpha 值，如#000000AA
fillColor	填充颜色	string	否	使用八位十六进制数表示，后两位表示 Alpha 值，如#000000AA
radius	半径	number	是	
strokeWidth	描边的宽度	number	否	

6.1.2 Map 组件控制 API

使用 wx.createMapContext(mapid)可以创建 Map 组件控制对象，该对象提供多个控制地图的方法。

- MapContext.getCenterLocation：获取当前地图中心的经纬度。返回 GCJ02 坐标系，可以用于 wx.openLocation。
- MapContext.moveToLocation(Object object)：将地图中心移至当前定位点，需要设置 Map 组件的 show-location 为 True。
- MapContext.translateMarker(Object object)：平移 marker，带动画。
- MapContext.includePoints(Object object)：缩放视野展示所有经纬度。
- MapContext.getRegion：获取当前地图的视野范围。
- MapContext.getRotate：获取当前地图的旋转角度。
- MapContext.getSkew：获取当前地图的倾斜角度。
- MapContext.getScale：获取当前地图的缩放级别。
- MapContext.setCenterOffset(Object object)：设置地图中心点偏移，向后、向下偏移为增长。屏幕比例范围为 0.25～0.75，默认偏移为[0.5, 0.5]。

6.1.3 Map 组件的使用

接下来创建 Map 组件小程序，先在页面加载 Map 组件，然后将地图定位在初始位置。

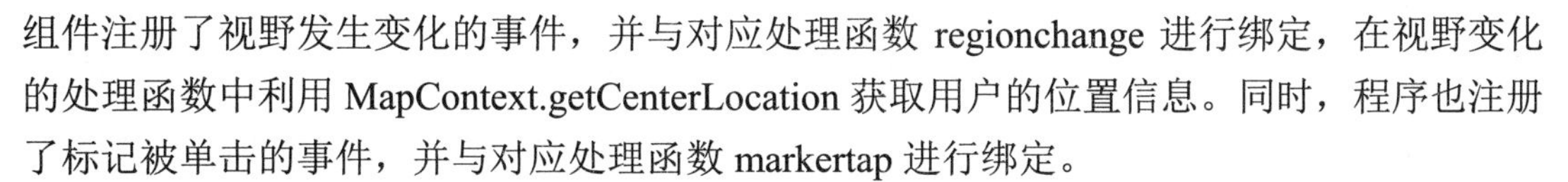

组件注册了视野发生变化的事件，并与对应处理函数 regionchange 进行绑定，在视野变化的处理函数中利用 MapContext.getCenterLocation 获取用户的位置信息。同时，程序也注册了标记被单击的事件，并与对应处理函数 markertap 进行绑定。

打开微信开发者工具，使用测试账号新建一个微信小程序，不使用模板。打开 index.wxml 文件，删除原来的代码，输入以下代码：

```
<map id="myMap" longitude="{{current_longitude}}" latitude=
"{{current_latitude}}" scale="14"
  markers="{{markers}}" bindmarkertap="markertap"
 polyline="{{polyline}}" bindregionchange="regionchange"
 show- location style="width: 100%; height: 300px;">
</map>
```

map 标签定义了一个 Map 组件，设置 id 值为 myMap，同时设置当前地图中心的经度、纬度和地图的放大级别。地图上还有标记点 markers 和多段线 polyline。

打开 index.js 文件，删除原来的代码，输入以下代码：

```
Page({
  data: {
    current_latitude:23.099994,
    current_longitude:113.324520,
    mapCtx:null, //Map 组件对象
    markers: [{
      iconPath: "/images/resources/others.png",
      id: 0,
      latitude: 23.099994,
      longitude: 113.324520,
      width: 50,
      height: 50
    }],
    polyline: [{
      points: [{
        longitude: 113.3245211,
        latitude: 23.10229
      }, {
        longitude: 113.324520,
        latitude: 23.21229
      }],
      color:"#FF0000DD",
      width: 4,
      dottedLine: true
    }]
  },
  onLoad(){
    this.setData({
    //创建 Map 组件对象
```

```
30          mapCtx:wx.createMapContext('myMap')
31        })
32      },
33      regionchange(e) {
34        console.log("视野发生变化：") ;
35        this.data.mapCtx.getCenterLocation({    //获取当前地图中心点的经度、
纬度
36        success: function(res){
37          console.log("当前位置纬度："+res.longitude)
38          console.log("当前位置经度："+res.latitude) }
39        })
40      },
41      markertap(e) {
42        console.log("单击了"+e.markerId+"号标记")
43      }
44    })
```

第 6～13 行定义地图上一个标记点的位置、使用的图标和大小。第 14～25 行定义多段线中的两个点的坐标。第 33 行定义一个地图视野发生变化事件对应的处理函数 regionchange。第 41 行定义一个标记被单击事件对应的处理函数 markertap。

6.1.4 小程序位置信息 API

微信小程序针对用户位置方面的应用提供了多个 API，以便开发人员进行开发。下面了解这些 API 的属性和回调函数的用法。

（1）wx.getLocation API：获取当前位置，参数说明如表 6.1.4 所示。

表 6.1.4 wx.getLocation API 的参数说明

属性	类型	是否必填	说明
Type	string	否	返回 GPS 坐标，默认为 WGS84。GCJ02 返回可用于 wx.openLocation API 的坐标
Altitude	string	否	传入 True 会返回高度信息。由于获取高度需要较高精确度，因此会减慢接口返回速度
isHighAccuracy	boolean	否	开启高精度定位
highAccuracyExpireTime	number	否	高精度定位超时时间（单位：ms）。在指定时间内返回最高精度，该值在 3000ms 以上时高精度定位才有效
Success	function	是	接口调用成功的回调函数
Fail	function	否	接口调用失败的回调函数
Complete	function	否	接口调用结束的回调函数（调用成功、调用失败都会执行）

该接口成功调用时返回的参数说明如表 6.1.5 所示。

表 6.1.5　wx.getLocation API 成功调用时返回的参数说明

参数	说明
Latitude	纬度，浮点数，取值范围为-90～90，负数表示南纬
Longitude	经度，浮点数，取值范围为-180～180，负数表示西经
Speed	速度，浮点数，单位为 m/s
Accuracy	位置的精确度
Altitude	高度，单位为 m
verticalAccuracy	垂直精度，单位为 m（安卓系统无法获取，返回 0）
horizontalAccuracy	水平精度，单位为 m

（2）wx.chooseLocation API：选择位置，参数说明如表 6.1.6 所示。

表 6.1.6　wx.chooseLocation API 的参数说明

属性	类型	是否必填	说明
Latitude	number	否	目的地纬度
Longitude	number	否	目的地经度
Success	function	否	接口调用成功的回调函数
Fail	function	否	接口调用失败的回调函数
Complete	function	否	接口调用结束的回调函数（调用成功、调用失败都会执行）

该接口成功调用时返回的参数说明如表 6.1.7 所示。

表 6.1.7　wx.chooseLocation API 成功调用时返回的参数说明

参数	说明
Latitude	纬度，浮点数，取值范围为-90～90，负数表示南纬
Longitude	经度，浮点数，取值范围为-180～180，负数表示西经
Name	位置名称
Address	详细位置

（3）wx.openLocation API：打开位置，参数说明如表 6.1.8 所示。

表 6.1.8　wx. openLocation API 的参数说明

属性	类型	是否必填	说明
Latitude	number	是	纬度，取值范围为-90～90，负数表示南纬。使用 GCJ02 坐标系
Longitude	number	是	经度，取值范围为-180～180，负数表示西经。使用 GCJ02 坐标系
Scale	number	否	缩放比例，取值范围为 5～18，默认为 18
Name	string	否	位置名称
Address	string	否	位置的详细说明
Success	function	是	接口调用成功的回调函数
Fail	function	否	接口调用失败的回调函数
Complete	function	否	接口调用结束的回调函数（调用成功、调用失败都会执行）

6.1.5 位置信息 API 的应用

利用上述位置信息 API，我们做了一个可以打开新位置和选择位置的小程序。在该小程序中，利用 Map 组件打开一个默认位置的地图，组件注册了一个视野发生变化的事件，并绑定对应的处理函数 regionchange，在函数中利用 wx.getLocation API 来获取移动后的经纬度数据。当单击“打开新位置”按钮时，调用 wx.openLocation API 来打开新位置的地图；当单击“位置选择”按钮时，调用 wx.chooseLocation API 来显示多个具体位置，并输出用户选择的位置信息。

在该示例小程序中，因为需要使用获取位置和选择位置的权限，所以在 app.json 文件中插入以下代码：

```
1    "permission": {
2      "scope.userLocation": {
3        "desc": "你的位置信息将用于小程序位置接口的效果展示"
4      }
5    },
6    "requiredPrivateInfos": [
7      "getLocation",
8      "chooseLocation"
9    ]
```

在 requiredPrivateInfos 中申请了获取位置和选择位置的权限，permission 中的配置会弹出是否允许小程序获取用户位置的选择窗口。

要实现示例中的效果，可以打开微信开发者工具，使用测试账号新建一个微信小程序，不使用模板。打开 index.wxml 文件，删除原来的代码，输入以下代码：

```
1   <map id="map" longitude="{{current_longitude}}"
2   latitude="{{current_latitude}}" scale="14"
3   bindregionchange="regionchange" show- location style="width: 100%;
4   height: 300px;"></map>
5   <view><button type="primary" style="margin-top:10px"
6   bindtap="openLocation">打开新位置</button>
7   <button type="primary" style="margin-top:10px"
8   bindtap="chooseLocation">位置选择</button></view>
```

第 1～4 行定义了一个 Map 组件，并设置了中心点位置的经纬度及地图的缩放级别；第 3 行注册了地图视野发生变化时对应的处理函数 regionchange；第 5 行和第 6 行定义的 button 绑定了它的单击事件的处理函数 openLocation；第 7 行和第 8 行定义的 button 绑定了它的单击事件的处理函数 chooseLocation。

打开 index.js 文件，删除原来的代码，输入以下代码：

```
1   Page({
2     data: {
3       current_latitude:23.099994,
```

```
4         current_longitude:113.324520,
5       },
6       regionchange(e) {
7         console.log("视野发生变化：")
8         wx.getLocation({
9           type:'gcj02',
10          success:res=>{
11            console.log("当前经度："+res.latitude) ;
12            console.log("当前纬度："+res.longitude) ;
13          },
14          fail(f){
15             console.log(f) ;
16          }
17        })
18      },
19      chooseLocation(e){
20        wx.chooseLocation({
21          success: function(res){
22            console.log("你选择的位置是："+res.address);
23          },
24          fail(e){console.log(e)}
25        })
26      },
27      openLocation(e){
28        wx.openLocation({
29          latitude: 23.099004,
30          longitude: 113.111520,
31          scale: 18
32        })
33      }
34    })
```

第 6～18 行定义了地图视野发生变化时对应的处理函数 regionchange，第 20～25 行是 wx.chooseLocation API 的使用代码，第 28～32 行是 wx.openLocation API 的使用代码。

任务二　使用腾讯地图 API

6.2.1　腾讯地图 WebService API

腾讯地图 WebService API 是基于 HTTPS/HTTP 的数据接口。开发者可以使用任何客户端、服务器和开发语言，按照腾讯地图 WebService API 规范，按需构建 HTTPS 请求，并获取结果数据（目前支持通过 JSON/JSONP 方式返回）。

首先需要注册位置服务开发者，并创建服务调用 key（key 是调用 API 的身份标识，作为必填参数之一传递给 API），以便在小程序中应用腾讯地图 WebService API。

在地址栏输入 https://lbs.qq.com/dev/console/application/mine，显示如图 6.2.1 所示页面。

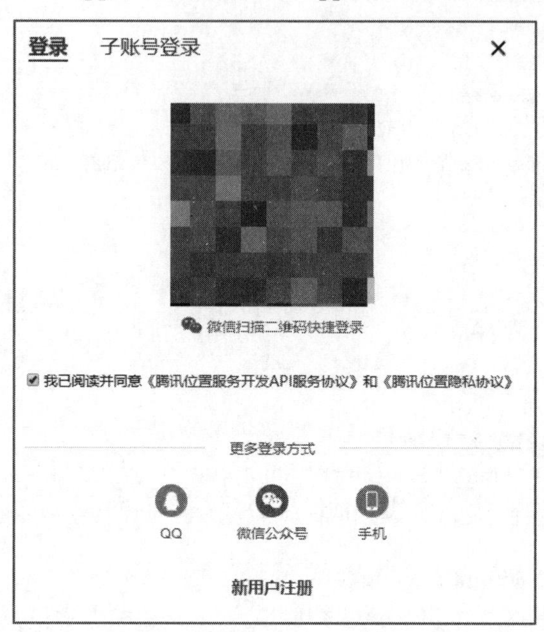

图 6.2.1　用户登录或注册页面

已经注册的用户使用微信扫描页面中的二维码进行登录，未注册的用户单击“新用户注册”按钮进行注册。在打开的“注册新账号”页面中填写手机号码、验证码和电子邮箱，如图 6.2.2 所示。

图 6.2.2　“注册新账号”页面

单击“绑定手机”按钮后登录，查看邮箱是否收到“腾讯位置服务”验证链接，如图 6.2.3 所示。

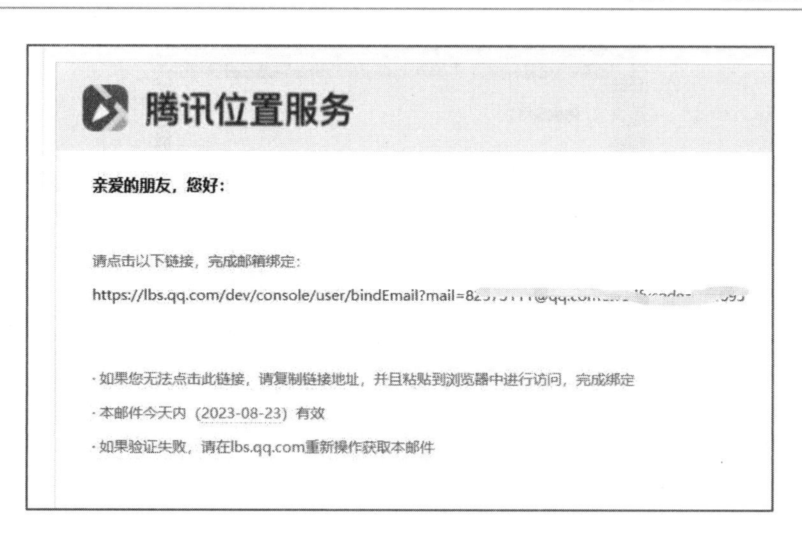

图 6.2.3　“腾讯位置服务”验证链接

单击链接后出现如图 6.2.4 所示的“注册成功”页面，单击“创建 API Key”链接，弹出如图 6.2.5 所示页面，单击“创建应用”按钮。

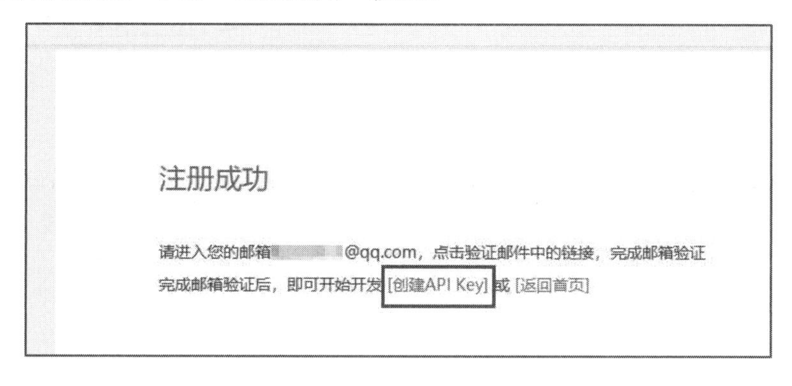

图 6.2.4　“注册成功”页面

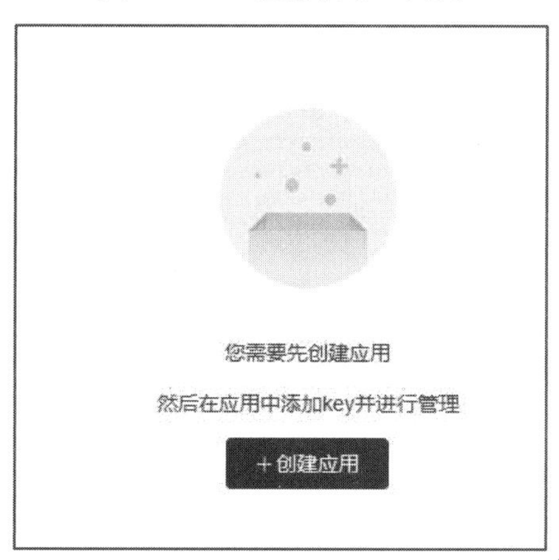

图 6.2.5　提示需要先创建应用页面

在弹出的如图 6.2.6 所示“创建应用”页面中填写应用名称和应用类型。

图 6.2.6　填写应用名称和应用类型

创建好应用后弹出如图 6.2.7 所示页面，单击“还没有 key，添加 key”链接。

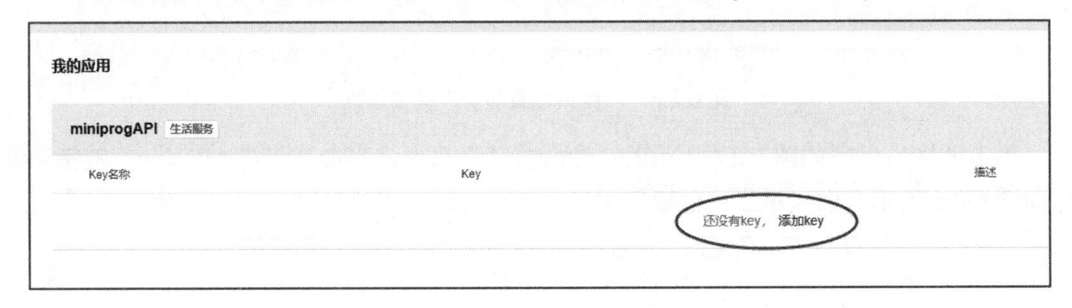

图 6.2.7　“我的应用”页面

在如图 6.2.8 所示页面中，勾选“WebServiceAPI”复选框，选中“域名白名单”单选按钮，填写“qq.com”和“servicewechat.com”等所有需要使用的域名，单击“保存”按钮。

Key名称* miniprogKey
描述 miniprog的key
启用产品* JavascriptAPI、JavascriptAPI GL、地图组件、Android定位SDK、iOS定位SDK在成功创建key后，即可使用
WebServiceAPI
域名白名单　授权IP　签名校验
适用于浏览器端JSONP方式调用Webservice服务
1.仅白名单中的域名才可使用此key调用WebService服务，留空则不限制
2.每行填写一个域名，填写的域名及其子域名都会同时得到授权，如：qq.com，则qq.com下的子域名map.qq.com也可以使用此key
qq.com
servicewechat.com
SDK
微信小程序
授权 APP ID* wx72006
取消　保存

图 6.2.8　添加 key 到应用的设置

在如图 6.2.9 所示页面中就出现我们需要的 key 了。

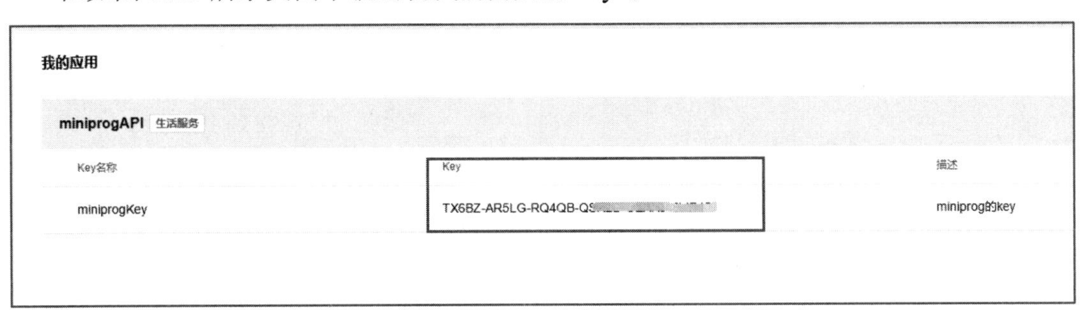

图 6.2.9　key 创建成功

6.2.2　小程序后台域名设置

应用的 key 申请完成之后，需要登录小程序后台进行设置，选择“开发管理-开发设置”选项，如图 6.2.10 所示。

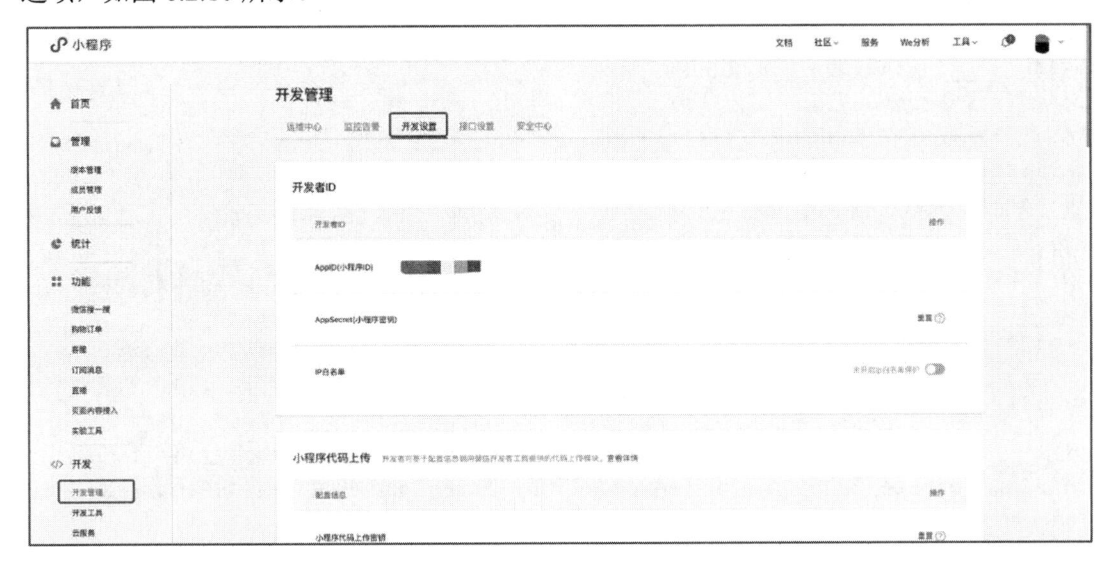

图 6.2.10　登录小程序后台进行设置

在“开发设置”页面进行服务器域名设置，在“服务器域名”中单击“修改”按钮，在 request 合法域名中填写“https://apis.map.qq.com”和“https://servicewechat.com”及其他需要使用的域名，单击“保存并提交”按钮。

6.2.3　地点搜索

创建完应用和 key，并且在小程序后台设置 request 的合法域名之后，就可以在小程序中应用腾讯地图 API 了。下面创建一个地点搜索小程序，打开微信开发者工具，使用测试账号新建一个微信小程序，不使用模板。打开 index.wxml 文件，删除原来的代码，输入以下代码：

```
<view class="container">
  <map id="map"
    class="map"
    markers="{{markers}}"
    longitude="{{longitude}}" latitude="{{latitude}}">
  </map>
</view>

<button size="mini" bindtap="buttonSearch">检索“美食”</button>

```

上述代码定义了一个 Map 组件，并设置了经纬度和标记点的属性。第 9 行创建了一个 button，并把它的单击事件与处理函数 buttonSearch 进行绑定。

打开 index.js 文件，删除原来的代码，输入以下代码：

```
Page({
  data: {
    latitude: 39.909088,
    longitude: 116.397643
  },

  buttonSearch(e){
    var _this = this
    var allMarkers = []
    //通过 wx.request 发起 HTTPS 接口请求
    wx.request({
      //腾讯地图 WebServiceAPI 地点搜索接口请求路径及参数（具体使用方法请参考开
      //发文档）
      url: 'https://apis.map.qq.com/ws/place/v1/search?page_index=
1&page_size=20&boundary=region(北京市,0)&keyword=美食&key=请填写你自己的
key',
      success(res){
        var pois = res.data.data;
        for(var i = 0; i< pois.length; i++){
          var title = pois[i].title
          var lat = pois[i].location.lat
          var lng = pois[i].location.lng
          var marker = {
            id: i,
            latitude: lat,
    longitude: lng,
    width:5,
    height:5,
            callout: {
```

```
27                    // 单击marker展示title
28                    content: title
29                  }
30                }
31                allMarkers.push(marker)
32                marker = null
33              }
34              _this.setData({
35                latitude: allMarkers[0].latitude,
36                longitude: allMarkers[0].longitude,
37                markers: allMarkers
38              })
39            }
40          })
41        }
42      })
```

在上述代码中，第 11 行调用了 wx.request 接口，并设置了 url 输入参数和 success 回调函数。在回调函数中，res.data.data 封装了所有搜索结果的地点信息。

搜索结果就是将附近的美食标注在地图上。

6.2.4　路径规划

除了可以进行地点搜索，腾讯地图 API 还可以进行路径规划。一个简单的示例小程序的 wxml 代码如下：

```
1     <view class="container">
2       <map id="map"
3         class="map"
4         polyline="{{polyline}}"
5         longitude="{{longitude}}" latitude="{{latitude}}" style="height:
      350px;">
6       </map>
7     </view>
8
9     <button size="mini" bindtap="buttonDriving">驾车路径规划</button>
```

上述代码定义了一个 Map 组件，并设置了经纬度和多段线 polyline 的属性。第 9 行创建了一个 button，并把它的单击事件与处理函数 buttonDriving 进行绑定。

路径规划示例小程序的 js 代码如下：

```
1     Page({
2       data: {
3         latitude: 39.909088,
4         longitude: 116.397643
```

```
5       },
6
7       buttonDriving(e){
8         var _this = this
9         //通过 wx.request 发起 HTTPS 接口请求
10        wx.request({
11          //腾讯地图 WebServiceAPI 驾车路径规划接口请求路径及参数（具体使用方法请参
            //考开发文档）
12          url: 'https://apis.map.qq.com/ws/direction/v1/driving/?key=你申
请的 key&from=39.894772,116.381668&to=39.932781,116.427171',
13          success(res){
14            var result = res.data.result
15            var route = result.routes[0]
16
17            var coors = route.polyline, pl = [];
18              //坐标解压（返回的点串坐标，通过前向差分进行压缩）
19              var kr = 1000000;
20              for (var i = 2; i < coors.length; i++) {
21                coors[i] = Number(coors[i - 2]) + Number(coors[i]) / kr;
22              }
23              //将解压后的坐标放入点串数组 pl 中
24              for (var i = 0; i < coors.length; i += 2) {
25                pl.push({ latitude: coors[i], longitude: coors[i + 1] })
26              }
27              _this.setData({
28                // 将路径的起点设置为地图中心点
29                latitude:pl[0].latitude,
30                longitude:pl[0].longitude,
31                // 绘制路径
32                polyline: [{
33                  points: pl,
34                  color: '#58c16c',
35                  width: 6,
36                  borderColor: '#2f693c',
37                  borderWidth: 1
38                }]
39              })
40          }
41        })
42      }
43    })
```

在上述代码中，第 10 行调用了 wx.request 接口，并设置了 url 输入参数和 success 回调函数。在回调函数中，res.data.result. routes[0].polyline 封装了规划路径上点的坐标信息。

运行结果就是从起点到终点绘制出一条规划路径。

任务三　在小程序中使用 Font Awesome 字体图标

6.3.1　使用 Font Awesome 字体图标准备工作

在小程序页面中使用 Font Awesome 的字体和图标，需要先进行以下准备工作。

首先，在 fontawesome 官方网站下载 font-awesome-4.7.0.zip 压缩包并解压。

然后，打开 transfonter 网站首页，单击“Add fonts”按钮后选择如图 6.3.1 所示的 ttf 文件，再单击“打开”按钮。

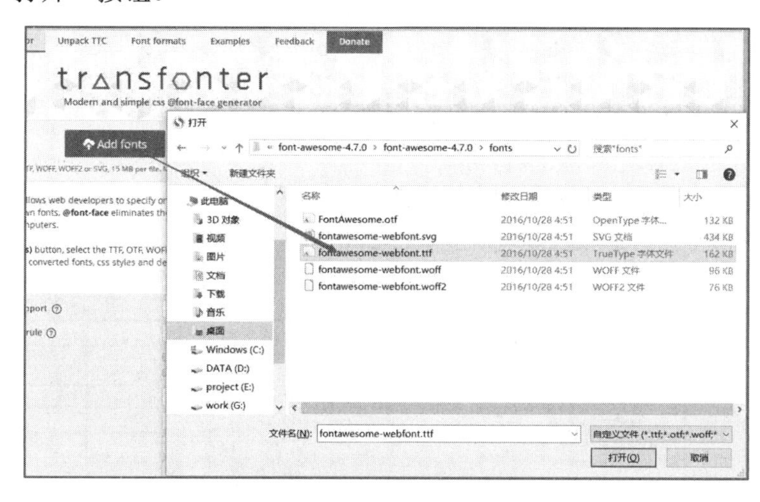

图 6.3.1　选择要转换的 ttf 文件

这一步操作选择的是解压后文件夹中的 fontawesome-webfont.ttf 文件。选择要转换的 ttf 文件后，按照如图 6.3.2 所示步骤进行操作：打开 Base64 encode 的开关→选中②中的全部复选框→单击“Convert”按钮。

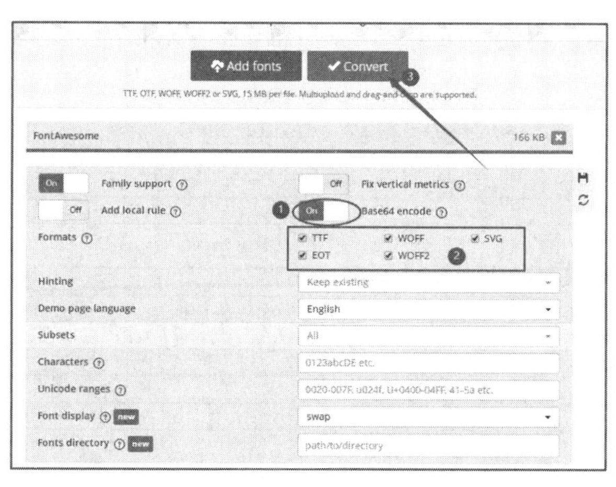

图 6.3.2　将 ttf 文件转换为 css 前的设置

转换完成之后，单击如图 6.3.3 所示的“Download”按钮，下载转换后的文件。

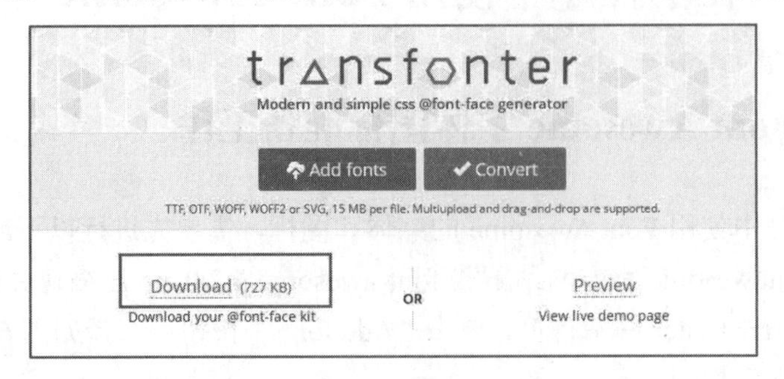

图 6.3.3　下载转换后的文件

6.3.2　应用 Font Awesome 字体图标

新建一个小程序项目，在项目文件夹下新建一个 font-awesome 文件。如图 6.3.4 所示，将从 6.3.1 节下载的文件 stylesheet.css 复制到 font-awesome 文件夹中，并重命名为 font-awesome-stylesheet.wxss。

› transfonter.org-20230822-145558

名称	修改日期	类型	大小
demo.html	2023/8/22 22:55	360 ChromeX H...	7 KB
FontAwesome.eot	2023/8/22 22:55	EOT 文件	162 KB
FontAwesome.svg	2023/8/22 22:55	SVG 文档	435 KB
FontAwesome.ttf	2023/8/22 22:55	TrueType 字体文件	162 KB
FontAwesome.woff	2023/8/22 22:55	WOFF 文件	96 KB
FontAwesome.woff2	2023/8/22 22:55	WOFF2 文件	75 KB
stylesheet.css	2023/8/22 22:55	层叠样式表文档	229 KB

图 6.3.4　复制并重命名下载的样式文件

复制从 font-awesome 解压出来的 font-awesome.css 到 font-awesome 文件夹中，并重命名为 font-awesome.wxss，将如图 6.3.5 所示方框中的@font-face 字体样式代码保存到 font-awesome.wxss 文件中。

```
1  /*!
2   *  Font Awesome 4.7.0 by @davegandy - http://fontawesome.io - @fontawesome
3   *  License - http://fontawesome.io/license (Font: SIL OFL 1.1, CSS: MIT License)
4   */
5  /* FONT PATH
6   * -------------------------
7  @font-face {
8    font-family: 'FontAwesome';
9    src: url('../fonts/fontawesome-webfont.eot?v=4.7.0');
10   src: url('../fonts/fontawesome-webfont.eot?#iefix&v=4.7.0') format('embedded-opentype'),
     url('../fonts/fontawesome-webfont.woff2?v=4.7.0') format('woff2'), url('../fonts/
     fontawesome-webfont.woff?v=4.7.0') format('woff'), url('../fonts/
     fontawesome-webfont.ttf?v=4.7.0') format('truetype'), url('../fonts/
     fontawesome-webfont.svg?v=4.7.0#fontawesomeregular') format('svg');
11   font-weight: normal;
12   font-style: normal;
13 } ------- */
14 .fa {
15   display: inline-block;
16   font: normal normal normal 14px/1 FontAwesome;
17   font-size: inherit;
18   text-rendering: auto;
19   -webkit-font-smoothing: antialiased;
20   -moz-osx-font-smoothing: grayscale;
21 }
22 /* makes the font 33% larger relative to the icon container */
23 .fa-lg {
```

图 6.3.5　@font-face 字体样式代码

修改 app.wxss 文件，如图 6.3.6 所示，在该文件中引入两个 wxss 文件并保存。

```
@import '/font-awesome/font-awesome-stylesheet.wxss';
@import '/font-awesome/font-awesome.wxss';

.container {
```

图 6.3.6　在 app.wxss 文件中引入两个 wxss 文件

在小程序的各个 wxml 页面中使用 font-awesome，如下列代码所示：

```
1  <i class="fa fa-circle-o-notch fa-spin"></i>
2  <i class="fa fa-refresh fa-spin"></i>
3  <i class="fa fa-cog fa-spin"></i>
4  <i class="fa fa-spinner fa-pulse"></i>
5
```

应用 font-awesome 的效果如图 6.3.7 所示。

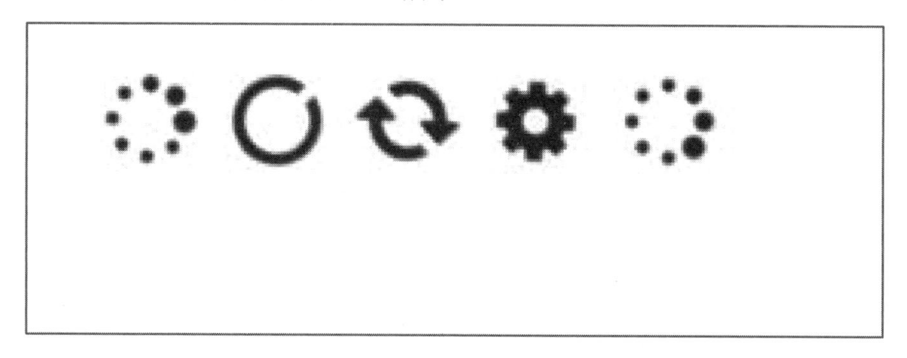

图 6.3.7　应用 font-awesome 的效果

任务四　制作半屏滑出效果

半屏滑出是一种页面交互方式，默认操作的半屏处于隐藏状态，当单击某个按钮或图标时显示操作的半屏；当单击“隐藏”按钮或背景空白处时，半屏滑出（隐藏）。要实现这种效果，需要页面布局 wxml、样式 wxss 和 js 三方面的配合。

打开微信开发者工具，使用测试账号新建一个微信小程序，不使用模板。打开 index.wxml 文件，删除原来的代码，输入以下代码：

```
1      <map class="map"  longitude="113.324520" latitude="23.099994"
scale="14" ></map>
2      <view class="route">
3        <image src="/images/路径.png" bindtap="showRouteText" />
4      </view>
```

```
5       <!-- 在遮罩层添加bindtap，可以在单击隐藏位置时隐藏dialog view -->
6       <view bindtap="showRouteText" class="mask {{show ? 'show':
  ''}}"></view>
7       <view class="dialog {{show ? 'show': ''}}">
8
9         <button type="default" bindtap="showRouteText" style=" position:
  fixed;bottom: 0px;">隐藏</button>
10      </view>
```

第1行生成底层地图，第2～4行生成一个图标层，单击图标使操作的半屏显示。第6行是一个遮罩层，单击遮罩层将操作的半屏隐藏。第7～10行是操作的半屏，单击其中的按钮可以隐藏操作的半屏。

wxss代码如下：

```
1     page{height: 100%;}
2     .map{width: 100%;height: 100%;}
3     .route {
4       position: fixed;
5       bottom: 10rpx;
6       right: 20rpx;
7     }
8     .route image {
9       width: 60rpx;
10      height: 60rpx;
11      border-radius: 50%;
12      border: solid rgb(62, 6, 214) 3rpx;
13    }
14     .mask {
15      position: fixed;
16      z-index: 1000;
17      top: 0;
18      right: 0;
19      left: 0;
20      bottom: 0;
21      /* 遮罩层背景 */
22      background: rgba(173, 16, 16, 0.6);
23      /* 遮罩层是隐藏状态 */
24      opacity: 0.5;
25      visibility: hidden;
26      /* 遮罩层动画，平移 */
27      transition: opacity .3s;
28    }
29
30    .mask.show {
31      /* 遮罩层是显示状态，顶部挡住所有的交互事件 */
32      opacity: 1;
```

```
33        visibility: visible;
34      }
35
36      .dialog {
37        position: fixed;
38        background-color: rgb(47, 204, 42);
39        left: 0;
40        right: 0;
41        bottom: 0;
42        min-height: 700rpx;
43        max-height: 75%;
44        z-index: 5000;
45        border-top-left-radius: 20rpx;
46        border-top-right-radius: 20rpx;
47        overflow: hidden;
48        transform: translateY(100%);
49        transition: transform .3s;
50      }
51
52      .dialog.show {
53        transform: translateY(0);
54        display: flex;
55        justify-content: center;
56      }
57
58      button {
59        margin: 10rpx;
60      }
```

第 14～18 行是遮罩层的默认样式，第 30～34 行是遮罩层显示时的样式，第 36～50 行是半屏的默认样式，第 52～56 行是半屏显示时的样式。

js 代码如下：

```
1     Page({
2       data: {
3         show: false
4       },
5       showRouteText: function (e) {
6         this.setData({ show: !this.data.show })
7       }
8     })
```

data 中的 show 变量用来控制遮罩层和操作半屏的显示或隐藏，第 5～7 行中的 showRouteText 函数被调用可改变 show 变量的状态值。

任务五　搜索餐厅、加油站等

6.5.1　项目介绍

本项目主要有地点搜索和路线规划两个功能。进入小程序来到用户当前位置（需要得到用户授权），地图顶部有六个按钮，分别是“餐厅”、“加油站”、“充电桩”、“地铁站”、“酒店”和“洗手间”，如图 6.5.1 所示。单击某个按钮搜索附近的地点，比如，单击“餐厅”按钮则搜索并标注附近的餐厅，单击“酒店”按钮则搜索并标注附近的酒店。

图 6.5.1　地点搜索及路线规划

单击某个餐厅或酒店等标注地点后，会显示从用户当前位置到此的路线规划，单击右下角的“路线”按钮，可以看到路线说明。单击“驾车”按钮，可以选择路线规划的方式，包括驾车、步行、骑车、电动车四种。任选一种路线规划方法，再单击标注地点，会重新进行路线规划。

6.5.2　项目初始化

打开微信开发者工具，新建项目，名称为 map，使用自己注册的 AppID，不使用模板。

打开 app.json 文件，将代码进行以下修改：

```
1 {
```

```
2      "permission": {
3        "scope.userLocation": {
4          "desc": "你的位置信息将用于小程序位置接口的效果展示"
5        }
6      },
7      "requiredPrivateInfos": [
8        "getLocation"
9      ],
10     "pages": [
11       "pages/index/index"
12     ],
13     "window": {
14       "backgroundTextStyle": "light",
15       "navigationBarBackgroundColor": "#fff",
16       "navigationBarTitleText": "地图",
17       "navigationBarTextStyle": "black"
18     },
19     "style": "v2",
20     "sitemapLocation": "sitemap.json"
21   }
```

复制资源文件夹中 map 下的 images 和 font-awesome 两个文件夹到新建的 map 微信小程序文件夹，并使其与 pages 文件夹并列。

6.5.3　注册腾讯地图 WebService API 开发者

本项目是通过调用腾讯地图 WebService API 来实现地点搜索和路线规划的。腾讯地图 WebService API 是基于 HTTPS/HTTP 的数据接口，在微信小程序中可以通过 wx.request 进行调用，还可以结合 Map 组件实现数据叠加展示、交互等应用需求。

打开浏览器，进入腾讯位置服务的首页，按照提示一步一步地注册。在申请腾讯位置服务 API Key 时，务必勾选“WebServiceAPI”复选框，如图 6.5.2 所示。

图 6.5.2　勾选“WebServiceAPI”复选框

注册成功后，保存腾讯位置服务 API 的 key，以备在代码中调用服务时使用。

6.5.4 实现地点搜索

微课：地图-地点搜索代码

打开 index.wxml 文件，删除原来的代码，输入以下代码：

```
1    <map id="mymap" bindmarkertap="markerTap" longitude="{{longitude}}" latitude="{{latitude}}" markers="{{markers}}" polyline="{{polyline}}" show-location></map>
2    <view class="poi">
3      <image wx:for="{{pois}}" wx:key="index" src="/images/{{item}}.png" id="{{index}}" bindtap="poiChange" />
4    </view>
```

第 2～4 行用于显示地图顶部的六个按钮。

打开 index.wxss 文件，输入以下代码：

```
1    page {
2      height: 100%;
3    }
4
5    map {
6      width: 100%;
7      height: 100%;
8    }
9
10   .poi {
11     position: fixed;
12     top: 10rpx;
13     left: 10rpx;
14     display: flex;
15   }
16
17   .poi image {
18     width: 60rpx;
19     height: 60rpx;
20     margin: 0 20rpx;
21     border-radius: 50%;
22     border: solid rgb(62, 6, 214) 3rpx;
23   }
```

打开 index.js 文件，删除原来的代码，输入以下代码：

```
1    const key = ''
2    let that
3    let map
4    let latitude = 23.13, longitude = 113.27, keyword = '餐厅'
```

```
5     let pois
6     Page({
7       data: {
8         longitude: 113.27, // 地图中心点的经度
9         latitude: 23.13, // 地图中心点的纬度
10        pois: ['餐厅', '加油站', '充电桩', '地铁站', '酒店', '洗手间'],
11      },
12
13      getLocation: function () {
14        //获取用户当前地理位置
15        //需要在app.json文件中配置permission和requiredPrivateInfos
16        wx.getLocation({
17          type: 'gcj02',
18          success: function (res) {
19            console.log(res)
20            latitude = res.latitude
21            longitude = res.longitude
22            that.setData({
23              latitude: res.latitude,
24              longitude: res.longitude
25            })
26          },
27          fail: err => {
28            console.log(err)
29          }
30        })
31      },
32
33      onLoad: function () {
34        that = this
35        map = wx.createMapContext('mymap')
36        this.getLocation()
37      },
38
39      search: function () {
40        var allMarkers = []
41        //通过wx.request发起HTTPS接口请求
42        wx.request({
43          //腾讯地图WebServiceAPI地点搜索接口请求路线及参数（具体使用方法请参考开
            //发文档）
          url: 'https://apis.map.qq.com/ws/place/v1/search?page_index=
44  1&page_size=20&boundary=nearby(' + latitude + ',' + longitude +
    ',1000,1)&keyword=' + keyword + '&key=' + key,
45          success(res) {
46            console.log(res.data)
47            pois = res.data.data
```

```
48            for (var i = 0; i < pois.length; i++) {
49              allMarkers.push({
50                id: Number(i),
51                iconPath: '/images/' + keyword + '.png',
52                width: 30,
53                height: 30,
54                latitude: pois[i].location.lat,
55                longitude: pois[i].location.lng,
56                callout: {
57                  // 单击marker展示title
58                  content: pois[i].title,
59                  display: 'ALWAYS',
60                  borderRadius: 5,
61                  borderWidth: 2
62                }
63              })
64            }
65            that.setData({
66              latitude: allMarkers[0].latitude,
67              longitude: allMarkers[0].longitude,
68              markers: allMarkers
69            })
70          },
71          fail: err => {
72            console.log(err)
73          }
74        })
75      },
76
77      poiChange: function (e) {
78        console.log(e)
79        keyword = this.data.pois[e.currentTarget.id]
80        map.getCenterLocation({
81          success: res => {
82            console.log(res)
83            latitude = res.latitude
84            longitude = res.longitude
85            this.search()
86          }
87        })
88      },
89
90    })
```

在第 1 行需要输入注册的 key。第 13 行中的 getLocation 方法用来获取用户当前地理位置。第 44 行按照腾讯地图 WebServiceAPI 地点搜索接口请求路线及参数生成 url，输入经纬度、关键字和注册的 key 进行地点搜索。第 48～75 行将接口返回的地点（包含经纬度）转换成 Map 组件的 allMarkers 数组，用于在地图上显示标记点。第 77 行的 poiChange 方法是单击地图顶部的六个按钮的事件处理方法，第 79 行根据单击的按钮得到对应的关键字，第 80～87 行调用 map.getCenterLocation 方法得到地图的中心点位置，第 85 行调用 search 方法进行地点搜索。

任务六　显示规划路线（驾车、步行、骑车、电动车）

扫一扫

微课：地图-路线规划 1

路线规划是地图的基本应用，方便了用户出行。本项目使用腾讯地图 WebService API 实现路线规划。

打开 index.wxml 文件，在后面输入以下代码：

```
1     <view class="route">
2       <image src="/images/路线.png" bindtap="showRouteText" />
3     </view>
4     <picker bindchange="pickerChange" value="{{mode_index}}" range=
  "{{modes}}">
5       <view class="picker">
6         <i class="fa {{modes_icon[mode_index]}}"></i> {{modes[mode_index]}}
7       </view>
8     </picker>
9
10    <!-- 在遮罩层添加bindtap，可以在单击隐藏位置时隐藏dialog view -->
11    <view bindtap="showRouteText" class="mask {{show ? 'show': ''}}">
  </view>
12    <view class="dialog {{show ? 'show': ''}}">
13      <scroll-view scroll-y>
14        <view class="routeText" wx:for="{{steps}}" wx:key="index">
15          <i class="fa {{modes_icon[mode_index]}}"></i> {{item.instruction}}
16        </view>
17      </scroll-view>
18      <button type="default" bindtap="showRouteText">隐藏</button>
19    </view>
```

第 6 行根据不同的路线规划模式，选择对应的 Font Awesome 字体图标。第 12～19 行实现半屏显示，结合样式和 js 代码可以实现半屏滑出效果。

扫一扫

微课：地图-路线规划 2

打开 index.wxss 文件，在开始处输入以下代码，用来访问 Font Awesome 字体图标。

```
1     @import '../../font-awesome/font-awesome.wxss';
2     @import '../../font-awesome/font-awesome-stylesheet.wxss';
```

在后面输入以下代码：

```
1   .route {
2     position: fixed;
3     bottom: 10rpx;
4     right: 20rpx;
5   }
6
7   .route image {
8     width: 60rpx;
9     height: 60rpx;
10    border-radius: 50%;
11    border: solid rgb(62, 6, 214) 3rpx;
12  }
13
14  picker {
15    position: fixed;
16    bottom: 20rpx;
17    right: 100rpx;
18    width: 130rpx;
19    border-radius: 10rpx;
20    border: solid rgb(62, 6, 214) 3rpx;
21    background: #eee;
22    padding: 10rpx;
23    text-align: center;
24    font-size: x-small;
25  }
26
27  .mask {
28    position: fixed;
29    z-index: 1000;
30    top: 0;
31    right: 0;
32    left: 0;
33    bottom: 0;
34    /* 遮罩层背景 */
35    background: rgba(0, 0, 0, .6);
36    /* 遮罩层是隐藏状态 */
37    opacity: 0;
38    visibility: hidden;
39    /* 遮罩层动画，平移 */
40    transition: opacity .3s;
41  }
42
43  .mask.show {
44    /* 遮罩层是显示状态，顶部挡住所有交互事件 */
45    opacity: 1;
```

```
46        visibility: visible;
47      }
48
49      .dialog {
50        position: fixed;
51        background-color: #fff;
52        left: 0;
53        right: 0;
54        bottom: 0;
55        min-height: 700rpx;
56        max-height: 75%;
57        z-index: 5000;
58        border-top-left-radius: 20rpx;
59        border-top-right-radius: 20rpx;
60        overflow: hidden;
61        transform: translateY(100%);
62        transition: transform .3s;
63      }
64
65      .dialog.show {
66        transform: translateY(0);
67      }
68
69      .routeText {
70        padding: 10rpx;
71        font-size: small;
72        text-align: center;
73      }
74
75      scroll-view {
76        height: 600rpx;
77      }
78
79      button {
80        margin: 10rpx;
81      }
```

打开 index.js 文件，在 data{}中插入以下代码：

```
1          modes: ['驾车', '步行', '骑车', '电动车'],
2          modes_icon: ['fa-car', 'fa-male', 'fa-bicycle', 'fa-motorcycle'],
3          steps: [],
4          mode_index: 0,
5          show: false
```

在 Page({})中补充几个方法，输入以下代码：

```
1        showRoute: function (route) {
```

```
2          console.log(route.steps)
3          this.setData({ steps: route.steps })
4          let coors = route.polyline, pl = [];
5          //坐标解压（返回的点串坐标，通过前向差分进行压缩）
6          const kr = 1000000;
7          for (var i = 2; i < coors.length; i++) {
8            coors[i] = Number(coors[i - 2]) + Number(coors[i]) / kr;
9          }
10         //将解压后的坐标放入点串数组 pl 中
11         for (var i = 0; i < coors.length; i += 2) {
12           pl.push({ latitude: coors[i], longitude: coors[i + 1] })
13         }
14         that.setData({
15           // 将路线的起点设置为地图中心点
16           latitude: pl[0].latitude,
17           longitude: pl[0].longitude,
18           // 绘制路线
19           polyline: [{
20             points: pl,
21             color: '#58c16c',
22             width: 6,
23             borderColor: '#2f693c',
24             borderWidth: 1
25           }]
26         })
27       },
28
29       // 使用腾讯地图 Direction API 进行路线规划
30       // to: 目的地（poi）
31       // 1. 驾车（driving）：支持结合实时路况、少收费、不走高速等多种偏好，精准预估
         // 到达时间（ETA）
32       // 2. 步行（walking）：基于步行的路线
33       // 3. 骑车（bicycling）：基于自行车的骑行路线
34       // 4. 电动车（ebicycling）：基于电动车的骑行路线
35       routePlan: function (to) {
36         const modes_en = ['driving', 'walking', 'bicycling', 'ebicycling']
37         const mode = modes_en[this.data.mode_index]
38         //获取用户当前地理位置
39         //需要在 app.json 文件中配置 permission 和 requiredPrivateInfos
40         //每次路线规划后都刷新当前位置
41         wx.getLocation({
42           type: 'gcj02',
43           success: function (res) {
44             console.log(res)
45             latitude = res.latitude
46             longitude = res.longitude
```

```
47             that.setData({
48               latitude: res.latitude,
49               longitude: res.longitude
50             })
51             wx.request({
52               //腾讯地图 WebServiceAPI 地点搜索接口请求路线及参数（具体使用方法请参
               //考开发文档）
53               url: 'https://apis.map.qq.com/ws/direction/v1/' + mode +
'/?from=' + latitude + ','
54                 + longitude + '&to=' + to.location.lat + ',' + to.
location.lng
55                 + '&to_poi=' + to.id + '&key=' + key,
56               success(res) {
57                 console.log(res.data)
58                 that.showRoute(res.data.result.routes[0])
59               },
60               fail: err => {
61                 console.log(err)
62               }
63             })
64           },
65           fail: err => {
66             console.log(err)
67           }
68         })
69       },
70
71       markerTap: function (e) {
72         console.log(pois[e.detail.markerId])
73         this.routePlan(pois[e.detail.markerId])
74       },
75
76       pickerChange: function (e) {
77         this.setData({
78           mode_index: e.detail.value
79         })
80       },
81
82       showRouteText: function (e) {
83         this.setData({ show: !this.data.show })
84       }
```

腾讯地图 WebService API 对返回路线规划的数据进行了压缩，第 6～9 行完成了解压缩，得到最终的经度或纬度。在第 11～13 行转换成经纬度坐标描述的点，存入数组 pl 中。在第 19～25 行转换成 Map 组件显示的折线，在地图上就可以看到规划好的路线了。

考虑到用户在使用小程序时可能移动位置，因此在第 41 行重新获取当前位置，并在第

51～63 行调用腾讯地图 WebService API 进行路线规划。第 53 行生成的 url 填入了规划模式（驾车、步行、骑车或电动车），当前位置的经纬度，目的地的经纬度、id 及 key。调用成功后，第 58 行调用 showRoute 来显示规划路线。

第 71 行中的 markerTap 是单击地图上的标记点触发的事件处理方法，只需调用前面的 routePlan 方法即可。

第 82 行中的 showRouteText 方法用来切换是否显示半屏的路线信息，当 show 为 true 时显示半屏，结合对应的布局和样式代码实现半屏滑出效果；当 show 为 false 时实现半屏隐藏。从 index.wxml 文件看出，showRouteText 方法在三个地方被调用。

（1）单击右下角的“路线”按钮。单击这个按钮会显示半屏滑出效果，因为半屏滑出后会遮住该按钮，因此该按钮仅实现半屏滑出显示效果。

（2）单击遮罩层。当显示半屏时，如果单击上半屏，遮罩层会触发 tap 事件，就会实现半屏隐藏。

（3）单击半屏中的“隐藏”按钮，也能实现半屏隐藏。

因为遮罩层和“隐藏”按钮只有半屏滑出后才能看到，因此单击遮罩层和“隐藏”按钮都仅实现半屏隐藏。

至此本项目完成，可以运行小程序进行测试。如果使用手机进行测试，则定位更精准。

项目小结

本项目主要学习了 Map 组件的基本用法，结合腾讯地图 WebService API，开发了地图应用中常用的地址搜索和路线规划功能。掌握这些基本技能后，学生可以根据本项目所学开发各种基于地图的微信小程序。本项目还介绍了 Font Awesome 字体图标的用法，以及常见的半屏滑出效果的实现方法。

习题

一、判断题

1．调用 wx.getLocation 可以获取用户的当前位置，但需要用户授权。（　　）

2．Map 组件支持地点搜索功能。（　　）

3．字体图标可以使用文字的方式显示。（　　）

二、选择题

1．设置 Map 组件的（　　）属性可以在地图上显示规划的路线。

A．markers

B．polyline

C．longitude

D．latitude

2．若想在地图上显示搜索到的位置，可以设置 Map 组件的（　　）属性。

A．markers

B．polyline

C．longitude

D．latitude

3．实现半屏滑出效果时，如果要显示遮罩层，则需要将 opacity 设置成（　　）。

A．0

B．0.5

C．0.8

D．1

三、填空题

1．腾讯地图 WebService API 是基于 HTTPS/HTTP 的数据接口，目前支持以________方式返回。

2．调用 MapContext.______________可以获取当前地图中心点位置。

3．要获取用户当前位置，需要在______________文件配置权限，获得用户授权。

四、编程题

扩展本项目案例，实现可以搜索附近的公园、博物馆、体育馆。

项目七　神秘的阴影

任务一　使用 Canvas 画图

与 HTML5 基本一致，在微信小程序中也使用 Canvas 来画图，画图的接口与 HTML5 的 Canvas 几乎一致。从基础库 2.9.0 开始，微信小程序支持一套新的 Canvas 2D 接口，原有的接口不再维护。新、旧接口差别不大，都需要使用画布上下文，旧接口使用 CanvasContext，新接口使用 RenderingContext，两个对象的使用方法非常接近。画布上下文的方法与 HTML5 中的基本一致，共有数十种，在此不一一介绍，读者可以参考 HTML5 中的画布上下文方法。

扫一扫

微课：神秘的阴影-画布

在微信小程序中使用 Canvas 画图的一般步骤如下。

（1）调用 wx.createSelectorQuery 方法获取布局文件中的 Canvas 组件。

（2）调用 canvas.getContext 方法得到画布上下文。

（3）调用画布上下文方法实现绘制。

①调用 clearRect 方法清屏。

②调用 beginPath 方法开始一段路线。

③调用 lineTo 方法画线，调用 arc 方法画圆或弧形。

④设置画线属性（如线宽 lineWidth、颜色 strokeStyle）。

⑤设置填充属性（如填充颜色 fillStyle）。

⑥调用 stroke 方法描边。

⑦调用 fill 方法填充。

下面通过制作一个自动旋转的五角星来说明在微信小程序中 Canvas 的用法。

打开微信开发者工具，使用测试账号新建项目，名称为 canvas，不使用模板。

打开 index.wxml 文件，删除原来的代码，输入以下代码：

```
1    <canvas type="2d" id="canvas" style="width: 100%; height: 100%;" />
```

指定 type="2d"，说明使用新的 Canvas 接口。

打开 index.wxss 文件，输入以下代码：

```
1    page {
2     height: 100%;
3    }
```

打开 index.js 文件，输入以下代码：

```
1    let canvas, ctx, dpr
2    let width, height
3    let timer
4    const horn = 5 // 绘制五个角
5    const angle = 360 / horn // 五个角的度数
6    // 两个圆的半径
7    let R, r
8    // 坐标
9    let x, y
10   // 旋转角度
11   let rotate = 0
12
13   function drawStar() {
14     // beginPath：开始绘制一段新的路线
15     if (ctx == null) return
16     ctx.clearRect(0, 0, width, height);
17     ctx.beginPath()
18     for (var i = 0; i < horn; i++) {
19       // 角度转弧度：角度/180*Math.PI
20       let rotateAngle = i * angle - rotate // 旋转起来
21       // 外圆顶点坐标
22       let a = rotateAngle / 180 * Math.PI
23       ctx.lineTo(Math.cos(a) * R + x, -Math.sin(a) * R + y)
24       // 内圆顶点坐标
25       let b = (rotateAngle + 360 / (2 * horn)) / 180 * Math.PI
26       ctx.lineTo(Math.cos(b) * r + x, -Math.sin(b) * r + y)
27     }
28     // closePath：关闭路线，将路线的终点与起点相连
29     ctx.closePath()
30     ctx.lineWidth = 3
31     ctx.fillStyle = '#E4EF00'
32     ctx.strokeStyle = "red"
33     ctx.fill()
34     ctx.stroke()
35     ctx.beginPath()
36     ctx.arc(x, y, R, 0, 2 * Math.PI)
37     ctx.lineWidth = 5
38     ctx.strokeStyle = 'red'
39     ctx.stroke()
40     rotate++
41   }
```

```
42
43  Page({
44    onLoad: function () {
45      // 通过 wx.createSelectorQuery 获取 Canvas 节点
46      wx.createSelectorQuery()
47        .select('#canvas')
48        .fields({
49          node: true,
50          size: true,
51        })
52        .exec(this.init.bind(this))
53    },
54
55    init(res) {
56      console.log(res)
57      canvas = res[0].node
58      ctx = canvas.getContext('2d')
59      dpr = wx.getSystemInfoSync().pixelRatio //手机像素比
60      width = res[0].width
61      height = res[0].height
62      canvas.width = width * dpr
63      canvas.height = height * dpr
64      R = width / 2 - 20
65      r = R / 2.7
66      x = width / 2
67      y = height / 2
68      ctx.scale(dpr, dpr) //设置缩放比例
69      drawStar()
70    },
71
72    onReady: function () {
73      timer = setInterval(drawStar, 30)
74    },
75
76    onUnload: function () {
77      clearInterval(timer)
78    }
79  })
```

第 46～52 行通过 wx.createSelectorQuery 获取 Canvas 节点，第 47 行中的#canvas 必须与 wxml 文件中 canvas 的 id 属性一致，否则无法获取 Canvas 节点。第 52 行通过绑定 init 方法来获取画布上下文等关键信息。第 58 行获取了画布上下文。第 59 行获取手机像素比，并结合第 68 行中的 scale 方法进行缩放，这样不论手机像素比为多少，都可以直接使用 res 参数的属性坐标值来绘图，不需要考虑缩放问题。比如，在第 60 行中得到的宽度值和在第 61 行中得到的高度值，可以直接用于绘图，不需要考虑因为手机像素比引起的缩放。在第

64 行中得到的 R 是五角星的外接圆半径，在第 65 行中得到的 r 是内部五个点的外接圆半径，第 66 行和第 67 行中的 x 和 y 是五角星的中心坐标。

创建完画布上下文，得到相关参数后，使用第 13 行中的 drawStar 方法绘制五角星。第 15 行清屏，第 17 行开始绘制一段新路线，第 18～27 行循环使用 lineTo 方法绘制五角星。第 29 行关闭路线。第 30 行设置线宽，第 31 行设置填充颜色，第 32 行设置线的颜色，第 33 行填充五角星，第 34 行对五角星描边。第 35～39 行用于绘制五角星的外接圆，第 36 行使用 arc 方法绘制圆，五个参数分别代表圆心 x 坐标、圆心 y 坐标、圆的半径、起始弧度、中止弧度（2PI 代表 360 度，即一个整圆）。

第 40 行中的 rotate 变量用于旋转五角星，初值在第 11 行设置为 0，在第 40 行进行累加，在第 18～27 行绘制五角星的代码中使用 rotate 来计算当前五角星的旋转角度。因此只需要反复调用 drawStar 方法，rotate 变量就会不断累加，从而实现五角星的旋转。

在第 73 行创建了一个定时器，每隔 30ms 执行一次 drawStar 方法。因此，小程序运行后，可以看到五角星旋转起来了。第 77 行清除定时器。

编译后，显示的页面如图 7.1.1 所示。

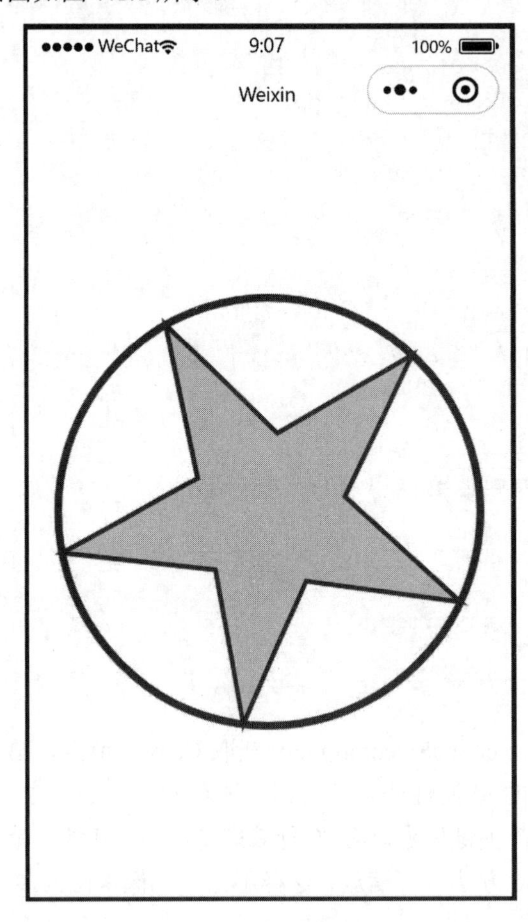

图 7.1.1　旋转的五角星

如果修改第 4 行中的 horn 常量，可以绘制其他 N 角星，图 7.1.2 所示是旋转的九角星。

图 7.1.2　旋转的九角星

任务二　制作神秘的阴影项目

7.2.1　项目介绍

本项目演示 Canvas 画布的用法，如图 7.2.1 所示。

屏幕中间有一个白色的光源，周围有多个大小不一的正方形盒子一边旋转一边移动，并且在光源的照射下拖着长长的影子。如果在手机上运行该项目，当倾斜手机时，盒子向倾斜的方向移动。在屏幕上单击或拖动会移动光源的位置。

图 7.2.1　神秘的阴影

扫一扫

微课：神秘的阴影-代码 1

7.2.2　制作方法

打开微信开发者工具，使用测试账号新建项目，名称为 shadow，不使用模板。打开 app.json 文件，修改“navigationBarTitleText”为“神秘的阴影”。

打开 index.wxml 文件，删除原来的代码，输入以下代码：

```
1      <canvas type="2d" id="canvas" style="width: 100%; height: 100%;"
   bindtouchstart="touchstart"  bindtouchmove="touchmove" />
```

为 canvas 设置了拖放事件 touchstart 和 touchmove 的处理方法，用来移动光源的位置，具体见 index.js 文件中的代码。

扫一扫

微课：神秘的阴影-代码 2

打开 index.wxss 文件，输入以下代码：

```
1     page {
2       height: 100%;
3     }
```

打开 index.js 文件，删除原来的代码，输入以下代码：

```
1   const minBoxSize = 5 //盒子的最小尺寸
2   let canvas, ctx, dpr
3   let width, height
4   let timer
5   //根据手机的倾斜方向确定 x 坐标和 y 坐标是增加还是减小，1 代表增加，-1 代表减小
6   //从而可以按照手机倾斜方向移动盒子
7   let xsign = 1, ysign = 1
8   let light = {
9     x: 160,
10    y: 200
11  }
12
13  //盒子的颜色
14  let colors = ["#f5c156", "#e6616b", "#5cd3ad", "#6ad34f", "#93f322"]
15
16  function drawLight() {
17    ctx.beginPath()
18    let m = width > height ? width : height
19    ctx.arc(light.x, light.y, m, 0, 2 * Math.PI)
20    //绘制放射状渐变
21    let gradient = ctx.createRadialGradient(light.x, light.y, 0,
light.x, light.y, m)
22    gradient.addColorStop(0, "#3b4654")
23    gradient.addColorStop(1, "#222")
24    ctx.fillStyle = gradient
25    ctx.fill()
26
27    ctx.beginPath()
28    ctx.arc(light.x, light.y, 10, 0, 2 * Math.PI)
29    gradient = ctx.createRadialGradient(light.x, light.y, 0, light.x,
light.y, 10)
30    gradient.addColorStop(0, "#fff")
31    gradient.addColorStop(1, "#3b4654")
32    ctx.fillStyle = gradient
33    ctx.fill()
34  }
35
36  function Box() {
37    this.half_size = Math.floor((Math.random() * 20) + minBoxSize)//外
接圆半径
38    this.x = Math.floor((Math.random() * width) + 1)
39    this.y = Math.floor((Math.random() * height) + 1)
40    this.r = Math.random() * Math.PI //旋转角度
41    this.shadow_length = 2000
```

```
42      this.color = colors[Math.floor((Math.random() * colors.length))]
43
44      this.getDots = function () {
45        let full = (Math.PI * 2) / 4
46        //盒子的四个顶点
47        let p1 = {
48          x: this.x + this.half_size * Math.sin(this.r),
49          y: this.y + this.half_size * Math.cos(this.r)
50        }
51        let p2 = {
52          x: this.x + this.half_size * Math.sin(this.r + full),
53          y: this.y + this.half_size * Math.cos(this.r + full)
54        }
55        let p3 = {
56          x: this.x + this.half_size * Math.sin(this.r + full * 2),
57          y: this.y + this.half_size * Math.cos(this.r + full * 2)
58        }
59        let p4 = {
60          x: this.x + this.half_size * Math.sin(this.r + full * 3),
61          y: this.y + this.half_size * Math.cos(this.r + full * 3)
62        }
63        return {
64          p1: p1,
65          p2: p2,
66          p3: p3,
67          p4: p4
68        }
69      }
70      //盒子的旋转速度和移动速度
71      this.rotate = function () {
72        let speed = (60 - this.half_size) / 10 //移动速度
73        this.r += speed * 0.01 //旋转速度
74        //xsign 和 ysign: 1 代表增加, -1 代表减小
75        this.x += speed * xsign
76        this.y += speed * ysign
77      }
78      this.draw = function () {
79        let dots = this.getDots()
80        ctx.beginPath()
81        ctx.moveTo(dots.p1.x, dots.p1.y)
82        ctx.lineTo(dots.p2.x, dots.p2.y)
83        ctx.lineTo(dots.p3.x, dots.p3.y)
84        ctx.lineTo(dots.p4.x, dots.p4.y)
85        ctx.fillStyle = this.color
86        ctx.fill()
87        //如果盒子移出了屏幕的下边界，则从屏幕的上边界出来
```

```
88          if (this.y - this.half_size > height && ysign == 1) {
89            this.y -= height + 100
90          }
91          //如果盒子移出了屏幕的上边界，则从屏幕的下边界出来
92          if (this.y - this.half_size < 0 && ysign == -1) {
93            this.y += height + 100
94          }
95          //如果盒子移出了屏幕的右边界，则从屏幕的左边界出来
96          if (this.x - this.half_size > width && xsign == 1) {
97            this.x -= width + 100
98          }
99          //如果盒子移出了屏幕的左边界，则从屏幕的右边界出来
100         if (this.x - this.half_size < 0 && xsign == -1) {
101           this.x += width + 100
102         }
103       }
104       this.drawShadow = function () {
105         let dots = this.getDots()
106         let points = []
107         for (let dot in dots) {
108           let angle = Math.atan2(light.y - dots[dot].y, light.x -
    dots[dot].x)
109           //计算该点的阴影终点坐标
110           let endX = dots[dot].x + this.shadow_length * Math.sin(-angle -
    Math.PI / 2)
111           let endY = dots[dot].y + this.shadow_length * Math.cos(-angle -
    Math.PI / 2)
112           points.push({
113             endX: endX,
114             endY: endY,
115             startX: dots[dot].x,
116             startY: dots[dot].y
117           })
118         }
119         for (let i = points.length - 1; i >= 0; i--) {
120           let n = i == 3 ? 0 : i + 1 //n是i的下一个点
121           ctx.beginPath()
122           //绘制相邻两个点和两个阴影终点组成的四边形
123           ctx.moveTo(points[i].startX, points[i].startY)
124           ctx.lineTo(points[n].startX, points[n].startY)
125           ctx.lineTo(points[n].endX, points[n].endY)
126           ctx.lineTo(points[i].endX, points[i].endY)
127           ctx.fillStyle = "#2c343f"
128           ctx.fill()
129         }
130       }
```

```
131     }
132
133     let boxes = []
134
135     function draw() {
136       if (!ctx) return
137       ctx.clearRect(0, 0, width, height)
138       drawLight()
139       for (let i = 0; i < boxes.length; i++) {
140         boxes[i].rotate()
141         boxes[i].drawShadow()
142       }
143       for (let i = 0; i < boxes.length; i++) {
144         //注释掉下面一行可以禁止碰撞检测，也可以避免盒子越来越小
145         //collisionDetection(i)
146         boxes[i].draw()
147       }
148     }
149
150     //碰撞检测
151     function collisionDetection(b) {
152       for (let i = boxes.length - 1; i >= 0; i--) {
153         if (i != b) {
154           let dx = (boxes[b].x + boxes[b].half_size) - (boxes[i].x +
boxes[i].half_size)
155           let dy = (boxes[b].y + boxes[b].half_size) - (boxes[i].y +
boxes[i].half_size)
156           let d = Math.sqrt(dx * dx + dy * dy)//两个盒子之间的距离
157           //如果盒子之间的距离太近，则减小盒子的尺寸，盒子会越来越小
158           if (d < boxes[b].half_size + boxes[i].half_size) {
159             boxes[b].half_size = boxes[b].half_size > minBoxSize ?
boxes[b].half_size -= 1 : minBoxSize
160             boxes[i].half_size = boxes[i].half_size > minBoxSize ?
boxes[i].half_size -= 1 : minBoxSize
161           }
162         }
163       }
164     }
165     Page({
166       onLoad: function () {
167         // 通过 SelectorQuery 获取 Canvas 节点
168         wx.createSelectorQuery()
169           .select('#canvas')
170           .fields({
171             node: true,
172             size: true,
```

```
173            })
174            .exec(this.init.bind(this))
175        },
176
177        init(res) {
178          console.log(res)
179          canvas = res[0].node
180          ctx = canvas.getContext('2d')
181          dpr = wx.getSystemInfoSync().pixelRatio //手机像素比
182          width = res[0].width
183          height = res[0].height
184          canvas.width = width * dpr
185          canvas.height = height * dpr
186          ctx.scale(dpr, dpr) //设置缩放比例，在绘图时可以参照width和height，如
   下面绘制圆的代码
187          // 绘制大圆
188          // ctx.clearRect(0,0,canvas.width,canvas.height)
189          // ctx.strokeStyle = '#ff0000'
190          // ctx.lineWidth = 1  // 设置线条的粗细，单位：px
191          // ctx.beginPath()    // 开始绘制一个新路线
192          // ctx.arc(width / 2, height / 2, width / 2, 0, 2 * Math.PI,
   true)
193          // ctx.stroke()
194          while (boxes.length < 18) {
195            boxes.push(new Box())
196          }
197          console.log(boxes)
198          draw()
199        },
200
201        onReady: function () {
202          timer = setInterval(draw, 30)
203          //使用手机加速度计，以判断手机的倾斜方向，确定x坐标和y坐标是增加还是减小
204          wx.onAccelerometerChange(function (res) {
205            //1代表增加，-1代表减小
206            xsign = res.x >= 0 ? 1 : -1;
207            ysign = res.y >= 0 ? -1 : 1;
208          })
209        },
210
211        setLightPos(e) {
212          if (e.touches.length != 1) return
213          light.x = e.touches[0].x
214          light.y = e.touches[0].y
215        },
216
```

```
217     touchstart: function (e) {
218       this.setLightPos(e)
219     },
220 
221     touchmove: function (e) {
222       this.setLightPos(e)
223     },
224 
225     onUnload: function () {
226       clearInterval(timer)
227     }
228   })
```

本段 js 代码比较复杂，下面进行详细介绍。

1. 绘制光源

第 16 行中的 drawLight 方法用于绘制光源。

小程序中的 Canvas 方法基本来源于 HTML5 中的 Canvas，比如第 21 行中的 ctx.createRadialGradient 与 HTML5 Canvas 中的相同，该方法创建放射状/圆形渐变对象。该方法的定义是 ctx.createRadialGradient(x0,y0,r0,x1,y1,r1)，六个参数分别代表渐变开始的 x 坐标、渐变开始的 y 坐标、开始圆的半径、渐变结束的 x 坐标、渐变结束的 y 坐标、结束圆的半径。以第 21 行代码为例，渐变开始于光源位置，开始圆的半径是 0，渐变结束于光源位置，结束圆的半径是 m（屏幕宽度和高度中的较大者），也就是从光源位置扩散渐变到整个屏幕。渐变颜色由第 22 行和第 23 行决定。第 21～23 行定义的放射状/圆形渐变用于绘制屏幕的背景，越远离光源，背景越暗，制造光源发光效果。类似地，第 29～31 行也定义了放射状/圆形渐变，用于绘制光源本身。

2. 定义盒子

第 36～131 行定义了盒子对象，其由若干属性和方法组成。第 37～42 行定义了六个属性，分别是盒子的外接圆半径、中心点 x 坐标、中心点 y 坐标、旋转角度、阴影长度和颜色。颜色来自第 14 行中的颜色数组，使用随机数产生下标，也就是随机选择颜色数组中的某种颜色。

后面定义了四个方法。

（1）第 44 行中的 getDots 方法获取盒子的四个顶点坐标，根据盒子中心点坐标、外接圆半径和旋转角度计算得来。

（2）第 71 行中的 rotate 方法完成盒子的旋转和移动。第 72 行根据盒子的外接圆半径确定移动速度，盒子越大移动速度越慢。第 75 行和第 76 行中的 xsign 和 ysign 代表 x 坐标和 y 坐标是增加还是减小，即确定盒子的移动方向，xsign 和 ysign 根据手机的倾斜方向确定，只有两个值：1 或-1。

（3）第 78 行中的 draw 方法用来绘制盒子。第 79～86 行绘制盒子。盒子在移动时可能

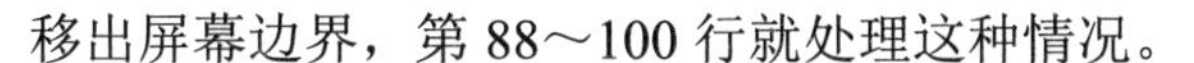

移出屏幕边界，第 88～100 行就处理这种情况。

（4）第 104 行中的 drawShadow 方法用来绘制阴影。第 107～118 行计算阴影的起点坐标和终点坐标。第 119～129 行通过绘制四边形的方式绘制阴影。

3．绘制所有图形

第 135 行中的 draw 方法用于绘制屏幕内的所有图形。第 137 行清屏。第 138 行绘制光源。第 139～142 行旋转、移动盒子并绘制所有阴影。第 143～147 行绘制所有盒子，其中，第 145 行用于碰撞检测，当盒子相撞时，通过减小盒子的尺寸来避免碰撞。可以注释掉该行，避免碰撞检测引起的盒子越来越小的情况。

4．碰撞检测

第 151 行用于碰撞检测。第 156 行计算盒子之间的距离。第 158 行判断盒子是否碰撞，如果距离 d 太小则判断为碰撞。第 159 行和第 160 行通过减小盒子的外接圆半径来避免碰撞。

5．创建 Canvas

第 168～175 行结合 init 方法来创建 Canvas。第 179 行得到 Canvas。第 180 行得到 Canvas Context。第 194～196 行创建 18 个盒子对象。

6．使用定时器

第 202 行定义了一个定时器，每隔 30ms 就调用 draw 方法去绘制屏幕，形成盒子移动效果及阴影的移动效果。第 226 行，在 onUnload 事件，即页面卸载时清除定时器。

7．使用加速度计获取手机倾斜方向

第 204 行中的 wx.onAccelerometerChange 监听加速度计数据事件。事件携带的 x、y、z 参数分别代表 3 轴的加速度值。第 206 行根据 x 值确定 xsign 值，即当 x>0 时，盒子向右移动（x 坐标递增），否则向左移动。第 207 行根据 y 值确定 ysign 值，即当 y>0 时，盒子向上移动（y 坐标递减），否则向下移动。

为什么 x>0 时，x 坐标递增，而 y>0 时，y 坐标递减呢？因为 x 坐标的正方向是屏幕向右，加速度计 x 轴的正方向也是屏幕向右；y 坐标的正方向是屏幕向下，但加速度计 y 轴的正方向是屏幕向上。可以看出 x 坐标和加速度计 x 轴的方向是一致的，而 y 坐标和加速度计 y 轴的方向是相反的。

8．处理拖放事件移动光源

第 217 行中的 touchstart 方法是手指触摸动作开始的事件处理方法，第 221 行是手指触摸后移动光源的事件处理方法，这两个方法都会调用第 211 行中的 setLightPos 方法。第 212 行检测是不是只有一根手指触摸，如果是两根以上手指触摸则不做处理，直接返回。根据第一根手指的坐标去设置光源的坐标，从而移动光源的位置。

至此，完成了本项目的所有代码，可以测试运行本项目。如果使用手机进行测试，可以查看手机倾斜后盒子的移动效果。

项目小结

本项目讲解在微信小程序中使用 Canvas 画图的方法，通过两个案例进行详细介绍。通过本项目的学习，可以掌握使用 Canvas 画线、填充的一般步骤，能够绘制常见的图形，实现动画效果。

习题

一、判断题

1．微信小程序中的 Canvas 画图使用了 HTML5 技术。（　　）

2．用画布上下文的 arc 方法可以绘制圆形。（　　）

3．用画布上下文的 strokeColor 属性设置描边的颜色。（　　）

二、选择题

1．获取页面布局文件中的 Canvas 组件，需要使用（　　）方法。

A．wx.getCanvas

B．wx.createSelectorQuery

C．wx.getSystemInfoSync

D．wx.select

2．画布上下文的 scale 方法的作用是（　　）。

A．设置坐标缩放比例

B．设置坐标旋转角度

C．设置坐标平移

D．设置坐标翻转

3．使用 Canvas 组件的新接口，需要将组件的 type 属性设置成（　　）。

A．d2　　　　B．d3

C．2d　　　　D．3d

三、填空题

1．调用画布上下文的____________方法填充，____________方法描边。

2．设置填充颜色，需要设置画布上下文的______________属性。

3．设置画线宽度，需要设置画布上下文的________________属性。

四、编程题

使用本项目学习的绘制五角星的方法，绘制中华人民共和国国旗（国旗的颜色、尺寸等参考《中华人民共和国国旗法》的“国旗制法说明”，可以查阅中华人民共和国中央人民政府网站的《中华人民共和国国旗法》页面。

项目八 手机助手

任务一 使用模板

如果一个微信小程序有很多页面，则通常有统一的页眉和页脚。如果使用复制代码的方式为每个页面分别添加页眉和页脚，一旦修改，就要修改每个页面，非常麻烦，也容易漏掉。这种情况下可以使用模板。

wxml 文件提供了模板功能，可以先在模板中定义代码片段，然后在不同的页面中调用。比如，可以定义页眉模板和页脚模板，并在需要的页面中进行调用。如果需要修改页眉或页脚，则只需要修改模板，不需要修改多个页面。

模板使用 name 属性作为名称，以方便调用。本项目在任务二中定义了页眉模板和页脚模板，此处不再举例。页眉模板的代码如下：

```
1    <template name="head">
2      <view class="page-head" wx:if="{{desc}}">{{desc}}</view>
3    </template>
```

页眉模板的名称为 head，页眉的内容可以通过传入变量 desc 来定制。调用模板使用 is 属性来指定模板名，调用页眉模板的代码如下：

```
<template is="head" data="{{desc: '页眉文字'}}" />
```

同样地，也可以定义页脚模板，代码如下：

```
1    <template name="foot">
2      <view class="page-foot">广东机电职业技术学院版权所有</view>
3    </template>
```

页脚模板的名称为 foot，调用页脚模板的代码如下：

```
<template is="foot" />
```

如果需要调整页眉、页脚的文字或样式，则只修改模板即可，所有调用页眉、页脚模板的页面会自动调整。

模板通过 import 关键字进行引用。假设页眉文件的完整路径是/lib/head.wxml，页脚文件的完整路径是/lib/foot.wxml，那么页面引用模板的代码如下：

```
1    <import src="../../lib/head.wxml" />
2    <import src="../../lib/foot.wxml" />
```

在 wxml 文件中，import 关键字用于引用模板，include 关键字用于引入 wxml 代码，相当于将 wxml 代码复制到 include 位置。

任务二　制作手机助手首页

8.2.1　项目介绍

本项目使用微信 API 访问手机的加速度计和指南针，实现了手机扫码及获取收货地址、发票抬头和手机系统信息等常见的手机助手功能，演示微信 API 提供的手机设备访问及常见信息的提取功能。

运行小程序后，首页如图 8.2.1 所示。

图 8.2.1　首页

单击“加速度计”显示如图 8.2.2 所示页面。

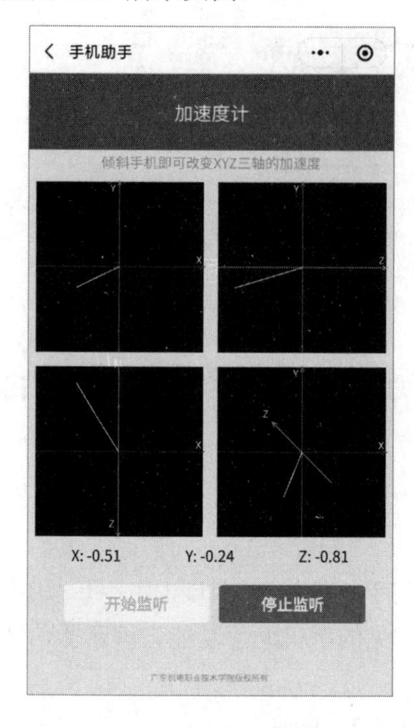

图 8.2.2　“加速度计”页面

单击“指南针”显示如图 8.2.3 所示页面。

图 8.2.3　“指南针”页面

单击“扫码”显示如图 8.2.4 所示页面，手机对准任何二维码或条码都可以扫码，图中显示了小程序的预览二维码。

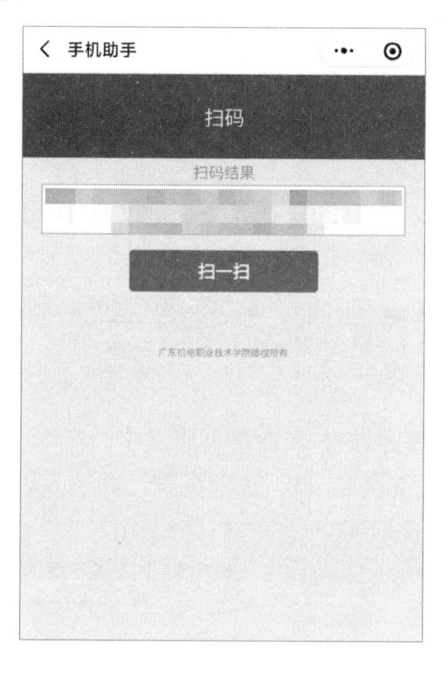

图 8.2.4　“扫码”页面

单击其他三个菜单项后的页面如图 8.2.5 所示（需要单击对应的按钮获取信息）。

图 8.2.5　获取收货地址、发票抬头和手机系统信息页面

8.2.2 制作模板

打开微信开发者工具，使用测试账号新建项目，名称为helper，不使用模板。打开app.json文件，修改“navigationBarTitleText”为“手机助手”。

复制资源文件夹中helper下的images文件夹到新建的helper微信小程序文件夹中，使该文件夹与pages文件夹并列。

在小程序根文件夹新建文件夹lib，在其下新建文件head.wxml，作为页眉，并输入以下代码：

```
1  <template name="head">
2    <view class="page-head" wx:if="{{desc}}">{{desc}}</view>
3  </template>
```

调用页眉模板时，通过指定desc来设置页眉文字。

在lib文件夹下新建文件foot.wxml，作为页脚，并输入以下代码：

```
1  <template name="foot">
2    <view class="page-foot">广东机电职业技术学院版权所有</view>
3  </template>
```

8.2.3 制作首页

扫一扫

微课：手机助手-首页

首先生成所有页面。打开app.json文件，按照以下代码进行修改。

```
1  {
2    "pages": [
3      "pages/index/index",
4      "pages/accelerometer/accelerometer",
5      "pages/compass/compass",
6      "pages/scan/scan",
7      "pages/address/address",
8      "pages/invoice/invoice"
9      "pages/getSystemInfo/getSystemInfo",
10   ],
11   "requiredPrivateInfos": [
12     "chooseAddress"
13   ],
14   "window": {
15     "backgroundTextStyle": "light",
16     "navigationBarBackgroundColor": "#fff",
17     "navigationBarTitleText": "手机助手",
18     "navigationBarTextStyle": "black"
19   },
20   "style": "v2",
```

```
21    "sitemapLocation": "sitemap.json"
22  }
```

第 4～9 行分别对应六个菜单项跳转的页面。按 Ctrl+S 组合键进行保存，或者单击工具条上的“编译”按钮将生成这六个页面。

调用 chooseAddress 方法需要在 app.json 文件中进行设置，设置方法见第 11～13 行。

接下来编写公共样式代码。打开 app.wxss 文件，删除原来的代码，输入以下代码：

```
1   page {
2     background: #eee;
3   }
4
5   .page-head {
6     padding: 50rpx 30rpx;
7     background: #07a307;
8     text-align: center;
9     color: #fff;
10    font-size: 40rpx;
11  }
12
13  .page-foot {
14    margin: 100rpx 0 30rpx 0;
15    text-align: center;
16    color: #aaa;
17    font-size: xx-small;
18  }
19
20  .page-section {
21    width: 100%;
22    margin-bottom: 60rpx;
23  }
24
25  .page-section_center {
26    display: flex;
27    flex-direction: column;
28    align-items: center;
29  }
30
31  .page-body-text {
32    font-size: 30rpx;
33    line-height: 26px;
34    color: #aaa;
35  }
36
37  .page-body-result {
38    min-height: 48rpx;
39    width: 90%;
```

```
40        text-align: center;
41        background: #fff;
42        word-break: break-all;
43        border: #aaa solid 1rpx;
44      }
45
46      button {
47        margin-top: 30rpx;
48      }
```

打开 index.wxml 文件，删除原来的代码，输入以下代码：

```
1     <import src="../../lib/head.wxml" />
2     <import src="../../lib/foot.wxml" />
3
4     <template is="head" data="{{desc: '手机助手'}}" />
5     <view class="menu">
6       <navigator wx:for="{{menu}}" wx:key="url" url="/pages/{{item.
url}}/{{item.url}}" class="navigator">
7         <view class="menu-text">{{item.text}}</view>
8         <view class="menu-arrow">
9           <image src="/images/arrow.png" />
10        </view>
11      </navigator>
12    </view>
13    <template is="foot" />
```

第 1 行和第 2 行分别引入了页眉模板和页脚模板。第 4 行调用了页眉模板，并指定标题为“手机助手”。第 13 行调用了页脚模板。本项目后面的每个页面都使用了页眉模板和页脚模板，不再赘述。

第 6～11 行中的 navigator 组件是页眉跳转组件，url 属性指定跳转的页眉 url。第 6～11 行还使用了列表渲染，将 js 代码中的 menu 数组转换成若干个 navigator 组件，以显示菜单项。

打开 index.wxss 文件，输入以下代码：

```
1     .menu {
2       margin: 10rpx;
3     }
4
5     .navigator {
6       display: flex;
7       align-items: center;
8       margin: 10rpx;
9       padding: 20rpx;
10      border-radius: 10rpx;
11      background: #fff;
12    }
13
```

```
14	.menu-text {
15	  flex: 1;
16	  color: #555;
17	}
18
19	.menu-arrow>image {
20	  width: 50rpx;
21	  height: 50rpx;
22	}
```

打开 index.js 文件，删除原来的代码，输入以下代码：

```
1	Page({
2	  data: {
3	    menu: [{
4	      url: 'accelerometer',
5	      text: '加速度计'
6	    }, {
7	      url: 'compass',
8	      text: '指南针'
9	    }, {
10	      url: 'scan',
11	      text: '扫码'
12	    }, {
13	      url: 'address',
14	      text: '获取收货地址'
15	    }, {
16	      url: 'invoice',
17	      text: '获取发票抬头'
18	    }, {
19	      url: 'getSystemInfo',
20	      text: '获取手机系统信息'
21	    }]
22	  },
23
24	})
```

在 data 中指定了 menu 数组，用于在首页中显示菜单项。

至此，首页制作完成，编译后可以看到首页中的菜单项。

任务三　使用手机加速度计

8.3.1　加速度计介绍

加速度计用来获取手机当前位置下三轴方向上的重力加速度分量，通过分量可以获取手机的倾斜方向。加速度计三轴的方向如图 8.3.1 所示。

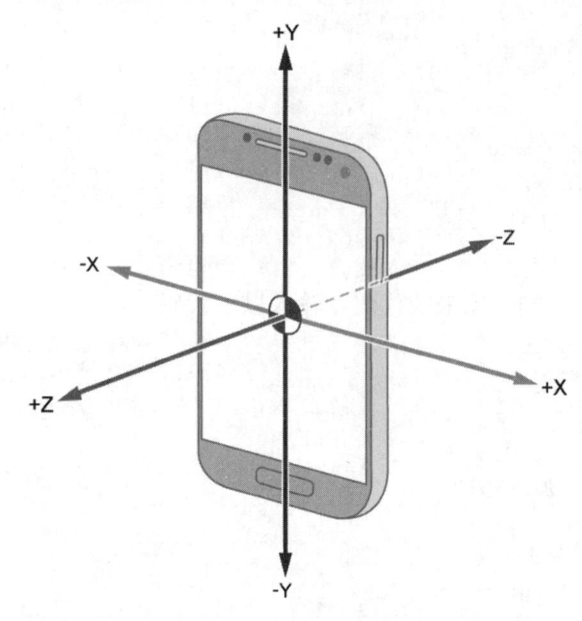

图 8.3.1　加速度计三轴的方向

在小程序中，有四个 API 方法用来处理加速度计，如表 8.3.1 所示。

表 8.3.1　API 方法

API 方法	说明
wx.onAccelerometerChange	监听加速度计数据事件。事件参数中的 x、y、z 分别代表三轴的加速度，取值范围是-1～1，乘以重力加速度即实际的加速度分量，方向如图 8.3.1 所示
wx.startAccelerometer	开始监听加速度数据。可以指定 interval 参数（类型是字符串）来设置监听加速度数据回调函数的执行频率，可选的设置如下。 game：适用于更新游戏的回调频率，在 20ms/次左右。 ui：适用于更新 UI 的回调频率，在 60ms/次左右。 normal：普通的回调频率，在 200ms/次左右
wx.stopAccelerometer	停止监听加速度数据
wx.offAccelerometerChange	移除加速度数据事件的监听函数

8.3.2　显示加速度计三轴数据

扫一扫

微课：手机助手-加速度计

打开 accelerometer.wxml 文件，输入以下代码：

```
1    <import src="../../lib/head.wxml" />
2    <import src="../../lib/foot.wxml" />
3
4
5    <template is="head" data="{{desc: '加速度计'}}" />
```

```
6
7      <view class="page-section page-section_center">
8        <text class="page-body-text">倾斜手机即可改变三轴的加速度</text>
9        <view class="page-body-canvas">
10         <canvas class="page-canvas" type="2d" id="canvas"></canvas>
11       </view>
12       <view class="page-body-xyz">
13         <text class="page-body-title">X: {{x}}</text>
14         <text class="page-body-title">Y: {{y}}</text>
15         <text class="page-body-title">Z: {{z}}</text>
16       </view>
17       <view class="page-body-controls">
18         <button type="primary" bindtap="startAccelerometer" disabled=
   "{{enabled}}">开始监听</button>
19         <button type="primary" bindtap="stopAccelerometer" disabled=
   "{{!enabled}}">停止监听</button>
20       </view>
21     </view>
22
23     <template is="foot" />
```

第 10 行嵌入了一个 canvas 组件，用于直观地显示加速度计数据。

打开 accelerometer.wxss 文件，输入以下代码：

```
1    .page-body-canvas {
2      width: 100vw;
3      height: 100vw;
4      position: relative;
5    }
6
7    .page-canvas {
8      position: absolute;
9      top: 0;
10     left: 0;
11     width: 100%;
12     height: 100%;
13   }
14
15   .page-body-xyz {
16     display: flex;
17     justify-content: space-between;
18     width: 700rpx;
19     box-sizing: border-box;
20     text-align: center;
21   }
22
23   .page-body-title {
```

```
24      margin-bottom: 0;
25      font-size: 32rpx;
26      width: 250rpx;
27    }
28
29    .page-body-controls {
30      margin-top: 10rpx;
31    }
32
33    .page-body-controls>button {
34      margin-left: 20rpx;
35      float: left;
36      width: 300rpx;
37    }
```

打开 accelerometer.js 文件，删除原来的代码，输入以下代码：

```
1   const margin = 8, qSqrt2 = 0.3536 //根号 2 的四分之一
2   let width, height
3   let canvas, ctx, dpr
4   let that
5   let x, y, z
6   Page({
7     data: {
8       x: 0,
9       y: 0,
10      z: 0,
11      enabled: true
12    },
13    onLoad: function () {
14      // 通过 SelectorQuery 获取 canvas 节点
15      wx.createSelectorQuery()
16        .select('#canvas')
17        .fields({
18          node: true,
19          size: true,
20        })
21        .exec(this.init.bind(this))
22    },
23
24    init(res) {
25      console.log(res)
26      canvas = res[0].node
27      ctx = canvas.getContext('2d')
28      dpr = wx.getSystemInfoSync().pixelRatio //手机像素比
29      width = res[0].width
30      height = res[0].height
```

```
31        canvas.width = width * dpr
32        canvas.height = height * dpr
33        ctx.scale(dpr, dpr)
34        this.draw()
35      },
36      onReady() {
37        that = this
38        //参数x、y、z，取值范围是-1~1
39        //X 轴向右是正方向，Y 轴向上是正方向，Z 轴垂直向上是正方向
40        wx.onAccelerometerChange(function (res) {
41          console.log(res)
42          x = res.x
43          y = res.y
44          z = res.z
45          that.setData({
46            x: res.x.toFixed(2),
47            y: res.y.toFixed(2),
48            z: res.z.toFixed(2)
49          })
50          that.draw()
51        })
52      },
53
54      //绘制向右的坐标轴
55      drawRightAxis(qwm, text) {
56        ctx.beginPath()
57        ctx.strokeStyle = text == 'X' ? '#ff0000' : '#00ffff'
58        ctx.moveTo(-qwm, 0)
59        ctx.lineTo(qwm, 0)
60        ctx.lineTo(qwm - 4, 2)
61        ctx.moveTo(qwm, 0)
62        ctx.lineTo(qwm - 4, -2)
63        ctx.fontSize = 20
64        ctx.stroke()
65        ctx.fillStyle = '#ffffff'
66        ctx.fillText(text, qwm - 8, -4)
67      },
68
69      //绘制向上的坐标轴
70      drawUpAxis(qwm, text) {
71        ctx.beginPath()
72        ctx.strokeStyle = '#00ff00'
73        ctx.moveTo(0, qwm)
74        ctx.lineTo(0, -qwm)
75        ctx.lineTo(2, 4 - qwm)
76        ctx.moveTo(0, -qwm)
```

```
77          ctx.lineTo(-2, 4 - qwm)
78          ctx.stroke()
79          ctx.fillText(text, -10, 10 - qwm)
80        },
81
82        //绘制向下的坐标轴
83        drawDownAxis(qwm, text) {
84          ctx.beginPath()
85          ctx.strokeStyle = '#00ffff'
86          ctx.moveTo(0, -qwm)
87          ctx.lineTo(0, qwm)
88          ctx.lineTo(2, qwm - 4)
89          ctx.moveTo(0, qwm)
90          ctx.lineTo(-2, qwm - 4)
91          ctx.stroke()
92          ctx.fillText(text, -10, qwm - 10)
93        },
94
95        //绘制呈 45 度的坐标轴
96        draw45Axis(qwm, text) {
97          let k = qSqrt2 * qwm
98          ctx.beginPath()
99          ctx.strokeStyle = '#00ffff'
100         ctx.moveTo(k, k)
101         ctx.lineTo(-k, -k)
102         ctx.lineTo(-k + 4 + 2, -k + 4 - 2)
103         ctx.moveTo(-k, -k)
104         ctx.lineTo(-k + 4 - 2, -k + 4 + 2)
105         ctx.stroke()
106         ctx.fillText(text, -k - 10, -k - 5)
107       },
108
109       draw() {
110         if (!ctx) return
111         let hw = width / 2 //宽度的二分之一
112         let qw = width / 4 //宽度的四分之一
113         let qwm = qw - margin //宽度的四分之一减去边界
114         let w = hw - 2 * margin //宽度的二分之一减去边界的两倍，是四个正方形的
    边长
115         //绘制左上角
116         ctx.fillStyle = '#000000'
117         ctx.translate(qw, qw)
118         ctx.fillRect(-qwm, -qwm, w, w)
119         this.drawRightAxis(qwm, 'X')
120         this.drawUpAxis(qwm, 'Y')
121         ctx.beginPath()
```

```
122         ctx.strokeStyle = '#ffffff'
123         ctx.moveTo(0, 0)
124         ctx.lineTo(x * qwm, -y * qwm)
125         ctx.stroke()
126         ctx.translate(-qw, -qw)//恢复
127 
128         //绘制右上角
129         ctx.fillStyle = '#000000'
130         ctx.translate(hw + qw, qw)//平移
131         ctx.fillRect(-qwm, -qwm, w, w)
132         this.drawRightAxis(qwm, 'Z')
133         this.drawUpAxis(qwm, 'Y')
134         ctx.beginPath()
135         ctx.strokeStyle = '#ffffff'
136         ctx.moveTo(0, 0)
137         ctx.lineTo(z * qwm, -y * qwm)
138         ctx.stroke()
139         ctx.translate(-hw - qw, -qw)//恢复
140 
141         //绘制左下角
142         ctx.fillStyle = '#000000'
143         ctx.translate(qw, hw + qw)//平移
144         ctx.fillRect(-qwm, -qwm, w, w)
145         this.drawRightAxis(qwm, 'X')
146         this.drawDownAxis(qwm, 'Z')
147         ctx.beginPath()
148         ctx.strokeStyle = '#ffffff'
149         ctx.moveTo(0, 0)
150         ctx.lineTo(x * qwm, z * qwm)
151         ctx.stroke()
152         ctx.translate(-qw, -hw - qw)//恢复
153 
154         //绘制右下角
155         ctx.fillStyle = '#000000'
156         ctx.translate(hw + qw, hw + qw)//平移
157         ctx.fillRect(-qwm, -qwm, w, w)
158         this.drawRightAxis(qwm, 'X')
159         this.drawUpAxis(qwm, 'Y')
160         this.draw45Axis(qwm, 'Z')
161         ctx.beginPath()
162         ctx.strokeStyle = '#ffffff'
163         ctx.moveTo(0, 0)
164         ctx.lineTo(x * qwm - z * qwm * qSqrt2, -y * qwm - z * qwm *
qSqrt2)
165         ctx.stroke()
166         ctx.translate(-hw - qw, -hw - qw)//恢复
```

```
167     },
168     startAccelerometer() {
169       if (this.data.enabled) {
170         return
171       }
172       const that = this
173       wx.startAccelerometer({
174         success() {
175           that.setData({
176             enabled: true
177           })
178         }
179       })
180     },
181     stopAccelerometer() {
182       if (!this.data.enabled) {
183         return
184       }
185       const that = this
186       wx.stopAccelerometer({
187         success() {
188           that.setData({
189             enabled: false
190           })
191         }
192       })
193     },
194   })
```

第 11 行中的 enable 变量用于控制页面上的“开始监听”和“停止监听”两个按钮是否变灰，从 accelerometer.wxml 文件可以看出，当 enable 为 true 时，“开始监听”按钮变灰；当 enable 为 false 时，“停止监听”按钮变灰。“开始监听”按钮对应的 tap 事件处理是第 168 行中的 startAccelerometer，第 173 行调用了 wx.startAccelerometer，在第 176 行将 enable 设置为 true；“停止监听”按钮对应的 tap 事件处理是第 181 行中的 stopAccelerometer，第 186 行调用了 wx.stopAccelerometer，在第 189 行将 enable 设置为 false。因此这两个按钮通过切换 enable 的值实现互锁。

canvas 的相关代码不再赘述，绘制坐标轴及对应三轴线段的代码请看代码中的注释，不再一一解释。

第 40～51 行调用 wx.onAccelerometerChange 读取加速度计数据，存放到全局变量 x、y、z 中，并通过调用 setData 刷新页面。因此页面会定时刷新，不需要使用定时器。

至此，完成了本任务的所有代码，使用手机进行测试，查看倾斜手机后的数据。

任务四　使用手机罗盘制作指南针

8.4.1　罗盘介绍

电子罗盘，也叫作数字指南针，通过测量地球磁场强度来确定当前的方位。现在电子罗盘广泛应用于手机等电子产品中。

在小程序中访问罗盘，主要有以下四种 API 方法，与访问加速度计类似，如表 8.4.1 所示。

表 8.4.1　API 方法

API 方法	说明
wx.onCompassChange	监听罗盘数据变化事件。频率：5 次/秒，接口调用后会自动开始监听，可使用 wx.stopCompass 方法停止监听。 事件参数 direction 代表面对的方向度数
wx.startCompass	开始监听罗盘数据
wx.stopCompass	停止监听罗盘数据
wx.offCompassChange	移除罗盘数据事件的监听函数

8.4.2　制作指南针

扫一扫

微课：手机助手-指南针

打开 compass.wxml 文件，删除原来的代码，输入以下代码：

```
1      <import src="../../lib/head.wxml" />
2      <import src="../../lib/foot.wxml" />
3
4
5      <template is="head" data="{{desc: '指南针'}}" />
6
7      <view class="page-body">
8        <view class="page-section page-section_center">
9          <text class="page-body-text">旋转手机即可获取方位信息</text>
10         <view class="direction">
11           <view class="bg-compass-line"></view>
12           <image class="bg-compass" src="/images/compass.png" style=
   "transform: rotate({{-direction}}deg)"></image>
13         </view>
14         <view class="direction-value">
15           <text>{{dir_text}}{{direction}}</text>
16           <text class="direction-degree">o</text>
17         </view>
18         <view class="controls">
```

```
19          <button type="primary" bindtap="startCompass" disabled=
    "{{enabled}}">开始监听</button>
20          <button type="primary" bindtap="stopCompass" disabled=
    "{{!enabled}}">停止监听</button>
21        </view>
22      </view>
23    </view>
24
25    <template is="foot" />
```

第 12 行中的“style="transform: rotate({{-direction}}deg)"”用于将图片旋转-direction 度，direction 是在 js 代码中读取的面对的方向度数。因为需要旋转图片，使顶部固定的指针指向对应的角度，所以图片旋转的角度与读取的度数应该方向相反，在其前面加上负号。

打开 compass.wxss 文件，输入以下代码：

```
1   .direction {
2     position: relative;
3     margin-top: 70rpx;
4     display: flex;
5     width: 540rpx;
6     height: 540rpx;
7     align-items: center;
8     justify-content: center;
9     background: #fff;
10    border-radius: 50%;
11    border: solid #6A4125 30rpx;
12  }
13
14  .direction-value {
15    position: relative;
16    font-size: 150rpx;
17    color: #6A4125;
18    line-height: 1;
19    z-index: 1;
20  }
21
22  .direction-degree {
23    position: absolute;
24    top: 0;
25    right: -40rpx;
26    font-size: 60rpx;
27  }
28
29  .bg-compass {
30    position: absolute;
31    top: 0;
```

```
32      left: 0;
33      width: 540rpx;
34      height: 540rpx;
35    }
36
37    .bg-compass-line {
38      position: absolute;
39      left: 267rpx;
40      top: -10rpx;
41      width: 6rpx;
42      height: 85rpx;
43      background-color: #e64211;
44      border-radius: 999rpx;
45      z-index: 1;
46    }
47
48    .controls {
49      margin-top: 70rpx;
50    }
51
52    .controls>button {
53      margin-left: 20rpx;
54      float: left;
55      width: 300rpx;
56    }
```

打开 compass.js 文件，删除原来的代码，输入以下代码：

```
1     Page({
2
3       data: {
4         enabled: true,
5         direction: 0,
6         dir_text: ''
7       },
8
9       onReady() {
10        const that = this
11        wx.onCompassChange(function (res) {
12          that.setData({
13            direction: res.direction.toFixed(0),
14            dir_text: that.get_dir_text(res.direction)
15          })
16        })
17      },
18      get_dir_text: function (direction) {
19        if (Math.abs(direction - 45) <= 22.5) return '东北'
```

```
20      if (Math.abs(direction - 90) <= 22.5) return '东'
21      if (Math.abs(direction - 135) <= 22.5) return '东南'
22      if (Math.abs(direction - 180) <= 22.5) return '南'
23      if (Math.abs(direction - 225) <= 22.5) return '西南'
24      if (Math.abs(direction - 270) <= 22.5) return '西'
25      if (Math.abs(direction - 315) <= 22.5) return '西北'
26      return '北'
27    },
28    startCompass() {
29      if (this.data.enabled) {
30        return
31      }
32      const that = this
33      wx.startCompass({
34        success() {
35          that.setData({
36            enabled: true
37          })
38        }
39      })
40    },
41    stopCompass() {
42      if (!this.data.enabled) {
43        return
44      }
45      const that = this
46      wx.stopCompass({
47        success() {
48          that.setData({
49            enabled: false
50          })
51        }
52      })
53    }
54  })
```

第 11 行调用 wx.onCompassChange 方法监听指南针数据，第 13 行将读取的指南针方向 direction 去掉小数部分，第 14 行通过 get_dir_text 方法获取对应的方位名称，如东、南、西、北、东南等。第 18 行中的 get_dir_text 方法将 360 度分成八等份，每份为 45 度，在八个方位的 45 度范围内对应各自的方位名称。比如，45 度对应东北，那么其附近的 45 度，即 22.5～67.5 度对应的方位就是东北。第 19 行使用 Math.abs 方法求数据的绝对值。

第 28 行中的 startCompass 方法和第 41 行中的 stopCompass 方法实现了“开始监听”和“停止监听”按钮的互锁，与加速度计的处理方法相同。

至此，完成了本任务的所有代码，使用手机进行测试，查看旋转手机后指南针的指向。

任务五　实现手机扫码

扫一扫

微课：手机助手-其他

大家经常会进行扫码支付、扫码点餐等，下面实现扫码功能。

打开 scan.wxml 文件，输入以下代码：

```
1   <import src="../../lib/head.wxml" />
2   <import src="../../lib/foot.wxml" />
3
4
5   <template is="head" data="{{desc: '扫码'}}" />
6
7   <view class="page-section page-section_center">
8     <view class="page-body-text">扫码结果</view>
9     <view class="page-body-result">{{result}}</view>
10    <button type="primary" bindtap="scanCode">扫一扫</button>
11  </view>
12
13  <template is="foot" />
```

本任务不需要编写样式文件代码，其包含在全局样式中。

打开 scan.js 文件，删除原来的代码，输入以下代码：

```
1   Page({
2     data: {
3       result: ''
4     },
5
6     scanCode() {
7       const that = this
8       wx.scanCode({
9         success(res) {
10          that.setData({
11            result: res.result
12          })
13        },
14        fail() { }
15      })
16    }
17  })
```

实现扫码功能非常简单，调用第 8 行中的 wx.scanCode 方法即可，第 11 行中返回的 result 参数即扫码结果。

至此，完成了本任务的所有代码，使用手机扫描任意二维码或条形码进行测试。

任务六　获取收货地址

开发商城、物流类的小程序需要获取用户的收货地址。wx.chooseAddress 用来实现获取收货地址的功能，需要在 app.json 文件中配置“"requiredPrivateInfos": ["chooseAddress"],”，在本项目的任务二中已经完成了配置。

打开 address.wxml 文件，删除原来的代码，输入以下代码：

```
1   <import src="../../lib/head.wxml" />
2   <import src="../../lib/foot.wxml" />
3
4
5   <template is="head" data="{{desc: '获取收货地址'}}" />
6
7   <view class="page-section page-section_center">
8     <view class="page-body-text">收货人姓名</view>
9     <view class="page-body-result">{{addressInfo.userName}}</view>
10    <view class="page-body-text">邮编</view>
11    <view class="page-body-result">{{addressInfo.postalCode}}</view>
12    <view class="page-body-text">地区</view>
13    <view class="page-body-result">
14      {{ addressInfo.provinceName }}
15      {{ addressInfo.cityName }}
16      {{ addressInfo.countyName }}
17    </view>
18    <view class="page-body-text">收货地址</view>
19    <view class="page-body-result">{{addressInfo.detailInfo}}</view>
20    <view class="page-body-text">国家码</view>
21    <view class="page-body-result">{{addressInfo.nationalCode}}</view>
22    <view class="page-body-text">手机号码</view>
23    <view class="page-body-result">{{addressInfo.telNumber}}</view>
24    <button type="primary" bindtap="chooseAddress">获取收货地址</button>
25  </view>
26
27  <template is="foot" />
```

本任务不需要编写样式文件代码，其包含在全局样式中。

打开 address.js 文件，删除原来的代码，输入以下代码：

```
1   Page({
2     data: {
3       addressInfo: null
4     },
5     chooseAddress() {
6       //需要在 app.json 文件中配置 "requiredPrivateInfos": ["chooseAddress"],
7       wx.chooseAddress({
```

```
8           success: (res) => {
9             this.setData({
10              addressInfo: res
11            })
12          },
13          fail(err) {
14            console.log(err)
15          }
16        })
17      }
18    })
```

wx.chooseAddress 方法使用非常简单，调用成功后返回地址信息，并保存在 addressInfo 中。之后直接在页面中显示即可。

至此，完成了本任务的所有代码，建议同时使用微信开发者工具和手机进行测试。

任务七　获取发票抬头

使用 wx.chooseInvoiceTitle 方法可以获取保存在微信中的发票抬头信息，包括抬头类型、抬头名称、抬头税号、单位地址等，方便开具发票。

打开 invoice.wxml 文件，删除原来的代码，输入以下代码：

```
1     <import src="../../lib/head.wxml" />
2     <import src="../../lib/foot.wxml" />
3
4     <template is="head" data="{{desc: '获取发票抬头'}}" />
5
6     <view class="page-section page-section_center">
7       <view class="page-body-text">抬头类型</view>
8       <view class="page-body-result">{{invoice.type !== '' ? (invoice.
  type === '0' ? '单位' : '个人') : ''}}</view>
9       <view class="page-body-text">抬头名称</view>
10      <view class="page-body-result">{{invoice.title}}</view>
11      <view class="page-body-text">抬头税号</view>
12      <view class="page-body-result">{{invoice.taxNumber}}</view>
13      <view class="page-body-text">单位地址</view>
14      <view class="page-body-result">{{invoice.companyAddress}}</view>
15      <view class="page-body-text">手机号码</view>
16      <view class="page-body-result">{{invoice.telephone}}</view>
17      <view class="page-body-text">银行名称</view>
18      <view class="page-body-result">{{invoice.bankName}}</view>
19      <view class="page-body-text">银行账号</view>
20      <view class="page-body-result">{{invoice.bankAccount}}</view>
```

```
21        <button type="primary" bindtap="chooseInvoiceTitle">获取发票抬头
   </button>
22      </view>
23
24      <template is="foot" />
```

本任务不需要编写样式文件代码，其包含在全局样式中。

打开 invoice.js 文件，删除原来的代码，输入以下代码：

```
1    Page({
2      data: {
3        invoice: null
4      },
5
6      chooseInvoiceTitle() {
7        wx.chooseInvoiceTitle({
8          success: (res) => {
9            this.setData({
10             invoice: res
11           })
12         },
13         fail: (err) => {
14           console.error(err)
15         }
16       })
17     }
18   })
```

wx.chooseInvoiceTitle 方法使用也非常简单，调用成功后返回发票信息，并保存在 invoice 中。之后直接在页面中显示即可。

至此，完成了本任务的所有代码，建议同时使用微信开发者工具和手机进行测试。

任务八　获取手机系统信息

wx.getSystemInfo 方法用来获取手机系统信息，包括手机品牌、手机型号、手机像素比等二十多种信息，详见微信官方文档。

打开 getSystemInfo.wxml 文件，删除原来的代码，输入以下代码：

```
1     <import src="../../lib/head.wxml" />
2     <import src="../../lib/foot.wxml" />
3
4     <template is="head" data="{{desc:'获取手机系统信息'}}" />
5
6     <view class="page-section page-section_center">
```

```
7         <view class="page-body-result">手机品牌：{{systemInfo.brand}}</view>
8         <view class="page-body-result">手机型号：{{systemInfo.model}}</view>
9         <view class="page-body-result">DPR：{{systemInfo.pixelRatio}}</view>
10        <view class="page-body-result">屏幕宽度：{{systemInfo.
   windowWidth}}</view>
11        <view class="page-body-result">屏幕高度：{{systemInfo.
   windowHeight}}</view>
12        <view class="page-body-result">微信语言：{{systemInfo.
   language}}</view>
13        <view class="page-body-result">微信版本：{{systemInfo.
   version}}</view>
14        <view class="page-body-result">操作系统及版本：{{systemInfo.
   system}}</view>
15        <view class="page-body-result">客户端平台：{{systemInfo.
   platform}}</view>
16        <button type="primary" bindtap="getSystemInfo">获取手机系统信息
   </button>
17      </view>
18
19      <template is="foot" />
```

打开 getSystemInfo.wxss 文件，输入以下代码：

```
1     .page-body-result {
2       margin: 10rpx 0;
3     }
```

打开 getSystemInfo.js 文件，删除原来的代码，输入以下代码：

```
1     Page({
2
3       data: {
4         systemInfo: {}
5       },
6       getSystemInfo() {
7         const that = this
8         wx.getSystemInfo({
9           success(res) {
10            that.setData({
11              systemInfo: res
12            })
13          }
14        })
15      }
16    })
```

wx.getSystemInfo 方法使用也非常简单，调用成功后返回手机系统信息，并保存在 systemInfo 中。之后直接在页面中显示即可。

至此，完成了本任务的所有代码，建议同时使用微信开发者工具和手机进行测试。

项目小结

本项目学习了模板、加速度计、指南针、扫码功能的实现，以及常见信息的查询方法，包括收货地址、发票抬头、系统信息等。通过学习，举一反三，能够正确使用手机的各种硬件并查询手机的各种信息。

习题

一、判断题

1．模板是 wxml 文件的一些代码片段，引入模板使用 include 指令。（　　）

2．手机加速度计的坐标方向与屏幕绘图的方向相同。（　　）

3．手机指南针监听事件参数 direction 代表面对的方向度数。（　　）

二、选择题

1．使用关键字（　　）调用指定名称的模板。

A．name

B．type

C．type

D．is

2．手机加速度计监听事件返回的 x、y、z 的取值范围是（　　）。

A．−1 到 1

B．−9.8 到 9.8

C．0 到 1

D．0 到 9.8

3．要获取手机型号，应该调用（　　）函数。

A．wx.scanCode

B、wx.chooseInvoiceTitle

C．wx.getSystemInfo

D．wx.getDeviceInfo

三、填空题

1．wx.onAccelerometerChange 是监听__________________数据事件。

2．wx.onCompassChange 是监听____________________数据事件。

3．微信小程序中的 wxml 文件提供了模板功能，使用模板的关键字是____________。

四、编程题

模仿本项目案例，通过调用 wx.getAppAuthorizeSetting 方法获取微信 App 授权设置。要求新建一个页面实现，并且在首页增加一个菜单项。